"A thoroughly enjoyable and mind-expanding array of puzzles and curiosities"
Dr Cliff Pickover, author of *Archimedes to Hawking*

"Clever, original brain-teasers are rare. This book has some beauties. Very satisfying."
Will Shortz, Crossword Editor, *The New York Times*

"I highly recommend this delightful book. It contains not only excellent puzzles, but also extremely interesting commentaries and anecdotes."
Professor Raymond Smullyan, author of *To Mock a Mockingbird*

"A wide-ranging and attractive collection that will appeal to all puzzle fans"
Professor Ian Stewart, author of *Professor Stewart's Cabinet of Mathematical Curiosities*

"A masterpiece ... reminds me somewhat of my first introduction to Martin Gardner ... this is a must for all puzzle lovers worldwide"
Terry Stickels, author of *Frame Games*

MATHEMATICAL PUZZLES

&
CURIOSITIES

Barry R. Clarke

DOVER PUBLICATIONS. INC
Mineola, New York

Acknowledgments

I should like to thank Denis Borris and Mark Rickert for testing some of the mathematics and logic puzzles in this work on the puzzles forum of my website http://barryispuzzled.com. Also, thanks to my many mathematics students for trying out some of the creative thinking puzzles and providing valuable feedback. I am grateful to Val Gilbert, and Alex and Kate Ware, for providing the opportunity to construct some of these puzzles for *The Daily Telegraph*. Finally, I am grateful to Rochelle Kronzek and James Miller from Dover Publications for their commitment to this work.

Bibliographical Note

Mathematical Puzzles and Curiosities is a new work, first published by Dover Publications, Inc., in 2013.

International Standard Book Number
ISBN-13: 978-0-486-49091-5
ISBN-10: 0-486-49091-2

Manufactured in the United States by Courier Corporation
49091203   2014
www.doverpublications.com

Contents

Introduction

First, here's a little enigma which at first sight seems trivial, but if you keep strictly to the given condition for solving it, then it's not so easy. Below are seven letters that form an anagram — and a seven-letter anagram is reasonably straightforward — but the puzzle is, how can you reach the solution *without* rearranging the letters? The answer, given at the end of the Introduction, demands a leap of the imagination but remember, if you succumb to the temptation to move the letters around then you've cheated!

LEtrECf

So, welcome, and I hope that you enjoy this original collection of puzzles and articles. It is a carefully considered compilation in which you can find conundrums in logic, mathematics, and creativity, together with some thought-provoking articles in recreational mathematics and philosophy. It is ideal for those who enjoy alternative ways of thinking and who like to consider a fresh approach to problems. Most of the book requires no specialised mathematical knowledge but those articles near the end that are more demanding can be penetrated by a reasonable amount of high-school algebra.

Some of the articles deal with classic teasers such as The Unexpected Hanging, The Monty Hall Problem, and The Sleeping Beauty Problem, but I have resisted resurrecting a standard analysis of these items, preferring instead to present my own way of understanding them. Other topics such as the Shakespeare Puzzles and Titan's Triangle are entirely new, the first being a creative interpretation of the dedications that preface the Shakespeare Sonnets (1609) and First Folio (1623), and the latter being a fascinating extension of a classic IQ puzzle. There are also more philosophical topics such as Zeno and Infinitesimals, and the Wave-Particle Puzzle, which I hope will encourage the reader to think again about these problems.

The puzzles in this work are entirely original and most of them have been published in my column in *The Daily Telegraph*. They have been arranged to increase in difficulty as the pages turn and the solutions have been deliberately placed out of order at the end to avoid inadvertently seeing the next solution. The answer to a puzzle can be located by referring to the solution number given at the end of the puzzle (not the page number) then looking it up at the back of the book. The creative thinking puzzles encourage alternative ways of thinking and two hints for each are provided near the end of the book to lessen the demand for mind reading. They are ideal for group problem-solving sessions and many of them have already been tested on students who have found them engaging, stimulating, and often amusing. If the solution remains beyond your grasp, even after pondering the hints, please don't feel frustrated. My wish for you is that on examining the answer you can enjoy it as puzzle art and, perhaps after having mastered a few of the basic principles, might even be inspired to create some of your own.

The following is an example of the kind of visual creative thinking puzzle that might be encountered within these covers. The solution is given at the end of the Introduction. Can you spot the difference between these two quarters?

 **spot the difference**

Having taught mathematics at various independent sixth-form colleges in Oxford for many years, one of my interests is in the role that puzzles, especially creative thinking ones, can play in developing the young mind. Sadly, the sole aim of our education system as it currently stands is entirely materialistic, this being to prepare students to obtain employment and earn a living. This usually requires qualifications such as a school certificate or university degree and our current school education system is one part of the long conveyor belt that serves this end. Of course, there are bills to pay so earning a living is far from being undesirable, but my point is that if this is the *only* goal then there is a price to pay.

When the teacher of mathematics presents his material, he does so with the intention of enabling his students to pass examinations. Mathematics examinations consist of questions and each question demands one or more methods to successfully negotiate it. If the student can identify the set of techniques required for each particular problem and accurately apply them then he can obtain a good result in the examination. The inquisitive student might express reservations about the usefulness of a particular problem to his future existence, ask for an insight into the history of the development of a certain topic, or even request a deeper understanding of why a method works, but none of this is vital

9

to achieving the eventual goal. To pass mathematics examinations, it is only necessary to understand the "how" and not the "why". A sympathetic teacher might wish to deviate from his rote-learning program and nurture the student's own capacity for independent thought, but with little time available to cover the curriculum it only compromises his own ability to deliver examination-successful students. Put another way, the current school system does not stimulate the student to think independently, and since most mathematics teachers enjoy exploring the background to their subject, I believe that given the choice they would prefer their students to do likewise.

This exclusive focus on examination results is a pity as far as the development of the student is concerned because there are a number of not only intellectual but also social advantages to be had by encouraging the student to explore and be creative. In the first place, a student who is engaged in creative activity is trying to invent possible ways of tackling a problem. This ability to think flexibly, which needs to be developed through repeated use, is vital to socially adaptive behaviour and leads to better social coping mechanisms. Also, a student who is trying out various possibilities to solve a problem, is bound to fail with some of the alternatives. So long as his educational environment is non-judgmental, it is possible for him to learn that this failure is a necessary part of problem solving, that it has a positive value in discovering what to eliminate from the enquiry, and that the next step is to try out other possibilities rather than fall into morbid introspection. The ability to take a risk in life, to fail, and then to be ready to try out a different approach, is a highly desirable skill.

It is unfortunate that the first real opportunity the student gets to truly explore original ideas usually only presents itself as late as doctorate level, by which time the young mind has been so deeply engraved with unquestionable 'truths' that it is almost impossible to think beyond them. Even then, some scientific PhD programs are so constrained they are nothing more than a passable permutation of the research department's existing ideas, accepted for publication in journals by devotees to the established order. There is nothing new in this circumstance, for it has existed since the birth of scientific enquiry in the Renaissance. The ideas of Aristotle had such a hold over university and church authorities in England in the late sixteenth century that it was deemed to be pointless asking questions apart from those intended to clarify Aristotle's work.

On a cultural level, the personalities who have advanced our civilisation the most are the ones who have refused to allow their creative instinct to be subdued by the rigours of rote and the strait-jacket of tradition. In science, Isaac Newton seemed not to enjoy the accepted Cambridge University curriculum, obtaining only an average undergraduate degree, but despite this he still managed to protect and develop his own ideas in mathematics and physics to make significant advances. Albert Einstein was so disinterested in rote learning that at the age of sixteen he failed his initial entrance examination for Zurich Polytechnic but still succeeded in bringing his ideas on relative motion in space and time to fruition, notions that had occupied him from the age of sixteen.

However, most students are not so fortunate and do not possess the unshakeable self-belief of a Newton or an Einstein. Years of rote learning eventually extinguishes their natural spirit of enquiry and it is no wonder that they become unable to creatively occupy themselves. The way forward is for teachers to find ways to nurture the inquisitive mind, to develop classes in creative thinking that encourage the student to try alternative approaches, support the student's inevitable failure, and celebrate his successes no matter how small.

Having said all this, creative thinking might not be for everyone ... but neither should it be stifled out of everyone. So for those who prefer to think for themselves, I hope that you enjoy the collection presented here and that it leads, if only in a small way, to new ways of considering problems.

Oh, by the way, the seven-letter anagram given earlier can be solved by placing a mirror above it at right angles to the page. The spot the difference puzzle needs a spot on the letter 'i' in the word 'difference'!

Happy thinking!

Barry R. Clarke
Oxford, UK, 2012
puzzledbarry@yahoo.co.uk

The Monty Hall Problem

On 9 September 1990, Marilyn vos Savant, puzzle columnist for the U.S. magazine *Parade*, published a probability teaser from one G. F. Whitaker, a reader from Columbia, Maryland. Now known as the Monty Hall Problem, the solution that she gave in her "Ask Marilyn" column produced enormous controversy resulting in over 10,000 mostly disagreeable letters. Apart from its challenge to common intuition, one reason the problem has attracted so much interest is that Ms vos Savant was once listed in the *Guinness Book of World Records* as having the highest recorded score in an IQ test (her 228 was based on a score attained at 10 years old). In fact, we shall see that although Ms vos Savant gave a definite solution, the conditions of the problem do not actually permit one.

Its earliest known form was known as Bertrand's Box Paradox[1] (1889) when the dilemma was stated as follows:

> Consider three boxes. One of these boxes contains two gold coins; one contains two silver coins; the other contains one silver coin and one gold coin. A box is chosen at random and one of the two coins inside is drawn. It turns out to be gold. What is the probability of the second coin in the box being gold?

At first glance, in order to have chosen a gold coin, then either box 1 or box 3 must have been chosen. Only one out of two of these boxes has a second gold coin so the probability is 1/2. However, let us label the gold and silver coins in the three boxes G_1 and G_2, S_1 and S_2, and S_3 and G_3, respectively. We are given that a gold coin has been taken first and so there are only three ways this could have happened: G_1G_2, G_2G_1 or G_3S_1. We can see that obtaining a second gold coin is possible in 2 out of the 3 cases and so the probability is 2/3.

A related teaser later appeared in Martin Gardner's *Mathematical Games* column from 1959 as The Three Prisoner Problem:[2]

> Of three prisoners, A, B, and C, one is to be executed and the other two set free. Prisoner A asks the warden which of B or C is to be released. The warden answers B? Does A or C now have the higher probability of being executed or are they equal?

	Executed	Released	Response
1	A	B,C	B
2	A	B,C	C
3	B	A,C	C
4	B	A,C	A→C
5	C	A,B	B
6	C	A,B	A→B

Let us construct a table of the six possible responses of the warden. In the first column is the prisoner to be executed and in the second we have the two prisoners to be released. Assuming the warden knows the identity of the victim, we use the third column to show his sometimes modified response (shown by the arrow) to the request "State the identity of a person other than A who is to be released". Suppose that this scenario is run every day for several years where each day one prisoner only is randomly chosen to be executed from three (labeled A, B, and C). Prisoner A is required to ask the same question of the warden each time. The six possible cases in the table will appear with almost the same relative frequency and whenever A hears the response "B" he knows that in only one out of three "B" responses A is to be executed whereas in two out of three "B" responses C is executed. So A knows that when the answer "B" is given by the warden his heaven-sent opportunity of execution occurs with a probability of 1/3 whereas for C it is 2/3.

In 1975 — inspired by Monty Hall's (real name Maurice Halprin) U.S. game show Let's Make a Deal — Steve Selvin sent two letters to *The American Statistician*.[3,4] The first contained the Monty Hall Problem more or less as it is now presented, and the second gave it its name. The puzzle, as presented by Ms vos Savant in 1990 (apart from the change in notation) ran as follows.

> Suppose you're on a game show and you're given the choice of three doors. Behind one is a car, behind each of the others is a goat. You pick a door, say door A, and the host, who knows what's behind the other doors, opens another door, say B, which has a goat. He then says: "Do you want to switch to door C?" Is it to your advantage to take the switch?

Despite the fact that the contestant has only doors A and C to choose from, and the probability of having the car seems to be 1/2 on switching, Ms vos Savant answered "yes" calculating the probability as 2/3. By way of explanation, she listed the three possible games arising from the contestant's selection of door A. Remember that the game show host has chosen, from B and C, a door with a goat and the result is for the contestant switching. So in two out of three games, the contestant wins a car if he switches from A to the unchosen door.

	Door A	Door B	Door C	Result
Game 1	car	goat	goat	lose
Game 2	goat	car	goat	win
Game 3	goat	goat	car	win

Those who judge the probability for this situation to be one half after the goat has been revealed, mistakenly treat the scenario as if the car has been randomly placed behind one of the other two doors: the contestant's original choice and the unopened door. The reality is that it has randomly been placed behind one of three doors, and that the contestant is given the choice between his own selection or the remaining pair, the latter having twice the probability of the former. When the game show host reduces the pair to one by revealing a goat, he doesn't alter the probability of the car being in that pair, and imparts no new knowledge to the contestant about whether or not the car is in that pair. It would be rather like the host declaring that a gold coin was hidden either inside a small trifle or a larger one ten times bigger. Then, knowing where the coin is located, the host carefully spoons away most of the large trifle, deliberately avoiding the coin (if it is there), thus reducing it to the size of the small trifle. However, reducing the size of the trifle does not alter the probability of it containing the coin, it only concentrates that probability into a smaller space.

Now here lies the difficulty. There is no indication in the problem as to the strategy employed by Monty the game show host, that is, whether or not: (i) he *always* reveals a goat behind one of the other two doors (i.e. his choice between B and C in Games 2 and 3 is governed by deliberate elimination); (ii) he sometimes reveals a goat and sometimes a car, even though he knows where the car is (i.e. he randomizes his choice between B and C); (iii) he only offers the switch when the contestant selects the prize door; or (iv) he only offers the switch when the contestant chooses a goat. This is important because it is his strategy that determines how the problem is analysed.

(i) Monty *always* reveals the goat. The probability tree diagram for this case is shown. We first examine how the probabilities have been arrived at after the contestant has chosen door A.

> **(a)** When the car is behind the contestant's choice, door A, we follow branch A on the left of the diagram by moving across to the right, and see that Monty cannot choose door A (probability 0) since the contestant has already chosen it, so he must open another door. Since a goat is behind each of B and C, Monty can randomly select which door to open (probability 1/2 each).

> **(b)** If the car is located behind door B, we follow branch B on the left of the diagram and find that Monty still cannot open the contestant's door A (probability 0), cannot reveal the car behind B (probability 0) and so only has door C to choose from (probability 1).

17

(c) The situation is symmetrical if the prize is behind C, for then Monty can open neither A (probability 0) nor C (probability 0) and only has the choice B (probability 1).

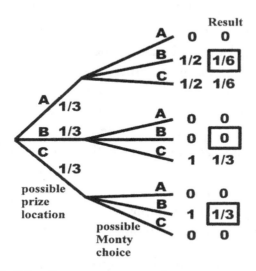

Probability diagram: contestant chooses door A, case (i).

The final column on the probability tree diagram shows the result of multiplying the probabilities horizontally to get the probability of combined events. For example, the probability that the car is behind B and Monty opens C is 1/3×1=1/3. In our problem, the contestant chooses door A and Monty opens B so we want to know the probability that C (the unselected door) has the car, given that Monty opens B. Now, all the probabilities relevant to Monty opening B are shown inside a square in the diagram.

These are the probabilities we can select from, and the particular one we are interested in is the C–B combination (bottom square) with probability 1/3 which contains C as the prize location. So the calculation is, *the probability that*

18

C has the car and Monty opens B (probability 1/3) divided by *the sum of the probabilities that are available to select from* designated by the square boxes, $1/6 + 0 + 1/3 = 1/2$. So the probability that C has the car when Monty opens door B is 2/3. A similar calculation for door A having the car when Monty opens B, gives 1/3 not 1/2 and so it is better to switch to door C which has probability 2/3. (In the game show Let's Make A Deal, Monty opened the door but never actually offered the switch.)

(ii) Monty *randomly* reveals a goat. The second option is that Monty randomly selects one of the remaining two doors B and C. As far as running a game show is concerned, running the risk of exposing the prize runs the risk of ruining the suspense. However, it is conceivable that Monty might do this if he decides that if he reveals the prize he will move items around and run the whole scenario again until a goat is revealed (and recorded TV has the advantage of only broadcasting this moment). It is irrelevant, however, to discuss degrees of reality; the point is, the problem does not prevent this. In this case, the goat he reveals behind door B might occur by chance and with probability 1/2. The next diagram shows how Monty's probabilities for the second set of branches are thus modified by following the given strategy.

The contestant has again chosen door A and we wish to calculate the probability that the car is behind the door C given that Monty has opened door B to reveal a goat (branch C–B). Noting that the branch B–B reveals the car, which is an invalid possibility, then the probabilities relevant to randomly revealing the goat are shown in only two square boxes and their sum is $1/6+1/6=1/3$. The prob-

19

ability that C–B occurs is 1/6. Division of the latter by the former now gives 1/2 which is the probability of the contestant being successful if he switches doors from A to C.

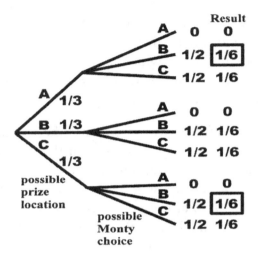

Probability diagram: contestant chooses door A, case (ii).

This illustrates the characteristic that originally produced the answer 2/3 for the switch to door C in (i). The fact that Monty had engaged in deliberate elimination for deciding which of the remaining two doors should be opened when a car might be revealed, transferred the total probability for the two doors onto the unopened door.

(iii) Contestant chooses prize door. If Monty offers the switch only when the contestant has correctly chosen the prize door then there is zero probability that the car is behind another door. Here, there is definitely no advantage to the contestant in switching. It would be a generous game show that consistently adopted this strategy, assuming the contestant was aware of it, because by offering the switch Monty reveals the location of the prize. However, the problem does not prevent this strategy.

(iv) Contestant chooses goat.

In this strategy, the switch is offered if the contestant does not choose the prize door and Monty always opens a door behind which is a goat. Provided the contestant is aware of this, once the game show host has revealed the goat, the contestant knows where the prize is.

Conclusion.

The correct solution to the Monty Hall Problem depends on one's interpretation of the way the game show is run in the problem. Some commentators only consider cases (i) and (ii) and say that the probability that the prize is behind C is at least 1/2 so one might as well switch. However, the problem asks "Is it to your *advantage* to take the switch?" and if strategy (ii) is adopted with the answer 1/2 then there is neither an advantage nor disadvantage. Our difficulty is, we simply do not know which strategy the game show host adopts, in other words, the possible strategies are not sufficiently narrowed for there to be a definite solution. An improvement on the Monty Hall problem has been given as follows:[5]

> A thoroughly honest game-show host has placed a car behind one of three doors. There is a goat behind each of the other doors. You have no prior knowledge that allows you to distinguish among the doors. "First you point toward a door," he says. "Then I'll open one of the other doors to reveal a goat. After I've shown you the goat, you make your final choice

whether to stick with your initial choice of doors, or to switch to the remaining door. You win whatever is behind the door." You begin by pointing to door number 1. The host shows you that door number 3 has a goat. Do the player's odds of getting the car increase by switching to door number 2?

Notice that the game show host declares unconditionally that he will *always* show a goat and *always* offer a switch, eliminating strategies (ii), (iii), and (iv). If the problem had been posed this way then Marilyn vos Savant would have provided the correct analysis.

References

1. Bertrand, Joseph. *Calcul de probabilities*, 1889.

2. Gardner, Martin. Mathematical Games column, *Scientific American* (October 1959), pp. 180–182. See also *The Second Scientific American Book of Mathematical Puzzles and Diversions*, Simon and Schuster: 1961.

3. Selvin, Steve. "A problem in probability" (letter to the editor), *The American Statistician*, 29(1):67 (February 1975).

4. Selvin, Steve. "A problem in probability" (letter to the editor), *The American Statistician*, 29(3):134 (August 1975).

5. Mueser, Peter R. and Granberg, Donald. "The Monty Hall dilemma revisited: understanding the interaction of problem definition and decision-taking." University of Missouri Working Paper 99-06, 1999.

Mathematical Puzzles I

1. The Baffled Brewer

Burper the brewer had just received an order for three barrels of beer, containing 16 litres, 10 litres, and 7 litres. He had the correct total quantity in three equal-sized barrels, however, the barrels were all half full. His problem was to produce the desired quantities in each barrel. To help him he had three measuring jugs having capacity three litres, two litres and one litre. Since Burper was clearly in a fix, he telephoned his mother on her cellphone who gave the following advice. Each measuring jug must be used once only to transfer a full jug of beer from one barrel to another. Furthermore, each barrel must be involved in at least two transfer operations so that on each occasion, there is either again or loss of beer. What operations are required?
[Solution 43]

2. Horse Play

At Dobbin's Dressage Academy, the 13 mounted horses were standing to attention around the hexagonal parade ground, with one in the centre. Each of the 13 riders had a number from 1–13, no two numbers appearing more than once, but only two riders, 4 and 12, were displaying them. Major Snodgrass, the instructor, had told them to arrange themselves so that each of the 12 straight lines of three numbers totalled 21. Can you fill in the missing numbers? [Solution 28]

3. Court Out

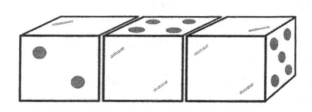

While the court was out, top barrister Betsy Gambols, Chancery Division, bet her adversary that he could not solve an enigma. Spotting that Betsy was attempting to obtain a psychological advantage, barrister Justin Case decided to face up to his opponent. The problem consists of three identical dice placed end to end in a line as shown. When the four missing numbers are correctly placed on the four blank faces, the numbers 1–6 will each appear once only on the three top and three front faces. The only restriction is that the total of the three front faces and the total of the four hidden inside faces are both even numbers. Can you avoid being caught out and place the four sets of spots on the four blank faces? [Solution 18]

4. Back to Class

Miss Flogginum, the mathematics teacher, was humiliating three of her pupils in front of the class. From left to right, Droopy, Dimwit and Dibdib stood with their back to the class and each had a digit from 0-9 pinned to his rear end, no digit being repeated. No pupil could see his own digit but was able to peep at the other two. The idea was that each had to make a statement about the two-digit number that would remain if he were not present. "It's the square of a whole number," said Droopy. "It's a prime number," reported Dimwit. Dibdib was bolder and surprised the class with a delightful observation. "The difference between the number and that formed by reversing the digits is three times the sum of the two digits." What digit did each pupil have? [Solution 40]

5. Sound Arithmetic

In Loudland, where everyone makes as much noise as possible, the inhabitants have three coins in their currency: a 3 honk, 4 honk, and 7 honk. One inhabitant recently bought a 2000 Watt stereo system for 63 honks using 15 coins, however, because of the noise going on around him, he couldn't recall how many of each coin type he had used. The three numbers of coins were each from 1 to 9 inclusive and no two numbers were the same. How many of each had he used? [Solution 44]

6. Digital Dilemma

Professor Crumble only had a good memory for digital facts. He once saw a five digit number which used some of the digits 1–9, no two digits being identical. He recalled that twice the third digit equalled the sum of the second and fourth ones, the fifth minus the second came to four, the third was a square number, the fifth was prime, and the fourth could be obtained by adding five to the first. Can you suggest where Professor Crumble saw this number? [Solution 1]

7. Core Conundrum

5	2	5	5
4	3	1	1
9	2	4	3
9	8	8	3

Old Mother Pip had a wooden box of apples partitioned into a 4×4 grid of compartments. Shown is the number of apples in each section. Rearrange the digits in the grid so that when completed, each row and column totals the same number. Only one row and one column is shown completely correct, with no other number correctly positioned. Can you find the correct arrangement? [Solution 14]

8. Word in the Stone

The archaeologist Doug Wither-Trowel was excavating a site in France one pleasant afternoon when he came upon an unusual stone with a four letter English word engraved upon it. In his notebook, he wrote that no two letters were identical and that if one replaced each letter in the word with a number giving its alphabetic position (A=1, B=2, etc), the total of the letters is 20. Not only that, but the sum of any three of the numbers is exactly divisible by the fourth number. When his French assistant read the entry in the notebook he came to the conclusion that the word was French. What was the English word on the stone? [Solution 5]

9. The Backward Robber

On Monday, a backward robber walked into a drugstore, pointed the gun at himself and handed the storekeeper half of the gold coins in his bag. The storekeeper, seeing his chance to make a handsome profit, demanded that the robber should also give him one third of the coins left in the bag. After counting out this number, the robber had a fit of belligerence and decided to give one half of the coins instead of one third.

Exactly the same thing happened on Tuesday, Wednesday, Thursday and Friday, the robber walking into the store with the same three-digit square number of coins in his bag. By the end of the week, the storekeeper had gained a cubic number of coins. How many coins did the storekeeper receive? [Solution 59]

10. Maximum Security

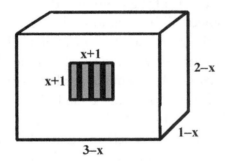

A prisoner requests a transfer to a cell with the maximum possible volume but unfortunately it is also the cell with the smallest possible window. All windows are $x+1$ square and all cells have dimensions $3-x$, $1-x$, and $2-x$, where x is a real number. What is the volume that no cell can exceed? [Solution 9]

Judgment Paradoxes

Hamlet and *Hales v Petit*

There is a scene in Shakespeare's *Hamlet* that appears to have borrowed from *Edward Plowden's Law Reports[1]* first published in 1571, and concerns the issue as to whether or not one can die in one's own lifetime. In 1554, Dame Margaret Hales brought an action against one Cyriac Petit who had been awarded her husband's estate, forfeited by the Crown after the death of her husband Sir James Hales, a Judge of the Court of the Common Pleas. His suicide by drowning followed his implication in the plot to place Lady Jane Grey — the granddaughter of Henry VIII's younger sister Mary Tudor — on the English throne. Apparently, his imprisonment and subsequent pardon so disturbed him that he walked into a river and ended his life. The Coroner returned a verdict of *felo de se*, that is, he was declared a felon against himself, a self-murderer, his guilt resting on the proof that he had the 'will and intention of committing it'. The punishments inflicted under English law were the forfeiture of a Christian burial and a transfer of all property to the Crown. In this case, the Crown intended to award Cyriac Petit the forfeited estate. The case centered on whether or not the grounds for forfeiture occurred during Sir James' lifetime, for if they had not, then his wife, Dame Margaret Hales, could preserve her entitlement.

Her counsel, Serjeants Southcote and Puttrel, argued that the termination of one's own life could not be completed in one's lifetime, because as long as one was still alive, the moment of death that would consummate the suicide had not yet been attained. A felony could only be committed by a person alive, but one would first have to lose one's life before a self-murder could be attributed, so a suicide should not be classed as a felony.

> He cannot be *felo de se* till the death is fully consummate, and the death precedes the felony and the forfeiture.

Serjeant Walsh, counsel for Petit the intended beneficiary, responded as follows:

> The act consists of three parts: the first is the imagination, which is a reflection or meditation of the mind, whether or not it is convenient for him to destroy himself, and what way it can be done; the second is the resolution, which is a determination of the mind to destroy himself; the third is the perfection, which is the execution of what the mind had resolved to do. And of all the parts, the doing of the act is the greatest in the judgment of our law, and it is in effect the whole.

Lord Dyer ruled that:

> the forfeiture shall have relation to *the act done* by Sir James Hales in his lifetime, which was the cause of his death, viz., the throwing himself into the water ... He therefore committed felony in his lifetime, although there was no possibility of the forfeiture being found in his lifetime, for until his death there was no cause of forfeiture.

Judge Browne delivered his own interpretation:

> Sir James Hales was dead. And how came he by his death? It may be answered, by drowning. And who drowned him? Sir James Hales. And when did he drown him? In his lifetime. So that Sir James Hales being alive caused Sir James Hales to die, and the act of the living was the death of the dead man. And for this offence it is reasonable to punish the living man, who committed the offence, and not the dead man.[2]

In *Hamlet*, 4.7.178–9, the Queen tells Laertes that his sister Ophelia, who has previously been showing symptoms of madness, has drowned. It appears not to have been deliberate, for she climbed a tree above a brook and the branch broke:

> *Queen.* … Which time she chanted snatches of old tunes,
> As one incapable of her own distresse …

In 5.1.1–22, two Clowns satirically discuss the drowning, echoing the discussion though not the outcome of the *Hales v Petit* case:

> *Clown.* Is she to be buried in Christian burial, that wilfully seekes her owne salvation?
> *Other.* … the Crowner hath sate on her, and finds it Christian burial.
> *Clown.* How can that be, unlesse she drowned her selfe in her owne defence?
> *Other.* Why, 'tis found so.
> *Clown.* It must be *Se offendendo;* it cannot bee else: for here lies the point; If I drowne myself wittingly, it argues an Act: and an Act hath three branches. It is an Act, to doe, and to perform; a*rgall* [therefore] she drown'd herself wittingly …
> Here lies the water; good: here stands the man; good: If the man goe to this water and drowne himself; it is will he nill he, he goes; marke you that? But if the water come to him & drowne him; he drownes not himself. *Argall* hee that is not guilty of his owne death shortens not his owne life.
> *Other.* But is this law?
> *Clown.* Ay' marry is't, Crowner's Quest Law.

As for the *Hales v Petit* case upon which this is based, a proof of intention is hazardous. If the felony is determined by the fact that Hales walked into the river and his death resulted, then every person who goes for a swim in the river and subsequently drowns is a felon. Surely the point is whether or not Hales had a 'determination of the mind' to drown himself, but that one has the intention of committing suicide cannot necessarily be assumed from the fact that death has resulted. There are many cases of a self-harming act that might have led to death where the subject has survived to protest it was only a 'cry for help.' It is safe to assume then that there are also cases where a 'cry for help' has resulted in death without there being an intention. Serjeants Southcote and Puttrel might have achieved a better outcome pressing this point.

The Case of Protagoras

According to Plato, the Greek philosopher Protagoras (490–420BC) was the first philosopher to teach virtue professionally. As well as being an agnostic, his interests also extended to politics and legal matters. There is a legend[3] that Protagoras struck a deal with one of his pupils so that the payment for his tuition was to be deferred until after he had won his first legal case. Unfortunately, on completing his training the student was dismayed to find that no clients were forthcoming. Even more dismayed was Protagoras who, being impatient, moved to sue his former pupil for the full amount outstanding with the confident declaration:

> Either I win this suit, or you win it. If I win, you pay me according to the judgment of the court. If you win, you pay me according to our agreement. In either case I am bound to be paid.

However, his adversary had a ready counter:

> If I win, then by the judgment of the court I need not pay you. If you win, then by our agreement I need not pay you. In either case I am bound not to have to pay you.

So whose argument was right?

Of course, Protagoras realizes that he has no legal grounds for bringing the case against his former student because at no time during the presentation of the arguments in court can the student be said to have already won his first case. So Protagoras knows that he must lose this argument about his former student reneging on their agreement and so is right in claiming that "If you win, you pay me according to our agreement", for the student would then have won his first case. His student is correct in thinking that "If I win, then by the judgment of the court I need not pay you" but only in the sense that at the time the arguments were presented he had not yet reneged on their agreement. However, after winning the case the agreement becomes active and he must now pay Protagoras who, if he brought the same case again would have a different argument and would certainly win. However, Protagoras's trick of forcing his student to win a case in court to receive payment might be subject to the penalty of having to pay costs when he loses. If this is so, then given that his former student fully intends to earn an income from his training then Protagoras's more profitable course might have been to wait.

Newcomb's Paradox

The following problem was devised by Dr William Newcomb of the Lawrence Livermore Laboratory, University of California but was first presented to the philosophy community by Robert Nozick.[4] A man is shown two closed boxes, A and B, on a table. It is known that A contains $1,000 whereas B contains either nothing or $1,000,000. No one knows which. The man is told that he can keep the contents of whichever boxes he chooses but only two possibilities are available to him. Either he can:

(1) Select both boxes.

(2) Select only B.

Prior to the test a superior intelligence from an alien civilization has predicted the man's choice. The Being's predictions are given to be "almost certainly" correct, resulting from a near perfect understanding of the man's thinking processes. If the Being has predicted that the man will choose both boxes then he has left B empty. However, if the Being predicts that the man will choose only B, then he has put $1,000,000 in it. Of course, the man might randomize his choice (e.g. by the toss of a coin) but in that case the Being will "almost certainly" have anticipated this and will have left B empty. In all cases A contains $1,000. These conditions are fully understood by everyone concerned. Everyone is also aware that everyone else understands them. How can the man maximize his winnings? Let us examine the consequences.

First, let us grant the premise that such a Being exists. It is evidently not a gainful course for the man to randomize his choice as the Being will "almost certainly" have predicted this and left box B empty. Since selecting box B is 'almost' necessarily connected with gaining $1,000,000 then the man might as well choose "B only". A counter argument has been given for taking both boxes, which develops as follows. At the point in time that the man is faced with his choice, box B either contains nothing or $1,000,000. The man's choice in the present cannot influence what the Being left in box B at some time in the past. So there is a chance that there is nothing in box B if the man chooses only that box. Surely, it would be better to take both boxes and be virtually guaranteed to receive at least $1000 and perhaps an extra $1,000,000. This is better than finishing with nothing at all.

However, no one is suggesting a reverse causality, only a forward causality where the man's choice in the present has been determined by the Being in the past and has influenced what the Being put in the boxes. In other words, whatever the man chooses has been calculated and the contents of the boxes have been decided by the Being on that basis. So if the Being is "almost certainly" correct then the man might as well choose box B. All objections to choosing "B only" must rely on doubt about the premises of the scenario. If we accept the predictive ability of the Being, then every single thought that the man entertains about the boxes has been accounted for by the Being. Since the choice "B only" necessitates a gain of $1,000,000 then this might as well be the selection. If we reject the assumption of the all-knowing Being then the problem is redundant.

In his 1969 paper, Nozick remarks that "To almost everyone, it is perfectly clear and obvious what should be done. The difficulty is that these people seem to divide almost evenly on the problem, with large numbers thinking that the opposing half is just being silly."

Newcomb's problem is evidently related to the issue as to whether or not free will exists. The superior Being in this problem could be interpreted as a completely determinable Nature. In this scenario, knowledge of the state of the local Universe in the present would then completely and mechanically decide all that occurs at a later moment including the choice from a set of possibilities that a human might make. This "state" would consist of the position and momentum of every elementary particle and photon that constitutes the decision-making part of the brain as well as information about all external objects that might affect it. Our possession of this knowledge of state would then allow us to make predictions about the choices made by the brain represented as a sophisticated electrical machine. There would be no free will.

While the Universe might operate in this way with all our choices predetermined, it will never be our privilege to view its operation. This limitation is provided by the nature of perception itself and of any measuring instrument that relies on being affected by the external world. It comes down to the Kantian distinction between the object outside of our minds that we cannot observe (Kant's "thing in itself") and the information (e.g. photons) that it sends to us as a change in itself. In fact, this is the upshot of naïve realism: we can never perceive the objects themselves but only the *changes* in these objects that are sent to us. So while we might be able to understand the mechanism of an object (e.g. a hydrogen atomic system) in the Universe — although present day quantum physicists seem to have abandoned this program — we will never be able to observe its operation and so we can never know its present "state" with the accuracy required to make a reliable prediction about aggregates of atomic changes that provide the electrical signals in the brain that constitute human choices. So, to those in fear of the loss of their free will, rest assured. The Universe might be evolving in a single predetermined course — although this is not the current academic view — but our incapacity to know what this is leaves the prediction of human choice to probability, not as the nature of knowledge, but as an approximation in the absence of it.

References

1. *Edward Plowden's Law Reports*, Part 1 (1571), pp.253-264.

2. Barton, Sir Dunbar Plunket, *Links Between Shakespeare and the Law* (Faber and Gwyer Ltd: 1929), pp.52-3.

3. Northrop, E.P., *Riddles in Mathematics* (Pelican: 1960), p.188.

4. Nozick, Robert (1969), "Newcomb's Problem and Two principles of Choice," in *Essays in Honor of Carl G. Hempel, ed. Nicholas Rescher*, Synthese Library (Dordrecht, the Netherlands: D. Reidel), p 115.

Finger Multiplication

The usual approach to teaching multiplication tables is by rote learning but, unfortunately, there is always a proportion of students who struggle to commit them to memory. Of course, there is the electronic calculator to assist but some examinations do not permit them and so any student who has not mastered these tables is at a disadvantage.

Here, a simple and original system is developed which, rather than relying on recall, allows students to calculate results of multiplications using fingers. Although applicable for tables 1 to 9, it turns out that it is easiest to use on the 7, 8 and 9 times tables, the ones that students traditionally have the greatest problem with. The task for the interested reader is to discover why the method works (an explanation is at the end of book).

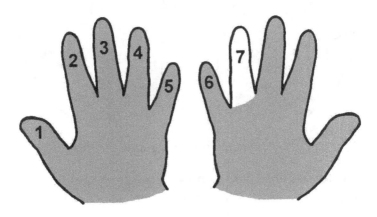

Calculation of 7×9 with nine times table.

Nine Times Table. As an example, we shall calculate the result of 7×9. Raise both hands and hold them together with your palms facing you. As with all calculations using this method, it will be important to remember the calculation position (here it is 7) in the chosen multiplication table. Now jumping one finger at a time, count seven from the left and drop the seventh finger (shown white). The result of the calculation has a tens digit and a units digit. The units digit is simply the number of fingers raised to the right of the dropped finger (3). The tens digit is the position in the multiplication table (7) minus the number of passes made through the ten fingers (1) to give $7-1 = 6$. So the result is 63. With regard to the number of passes, we note that for other multiplication tables below nine, the ten fingers can be traversed more than once.

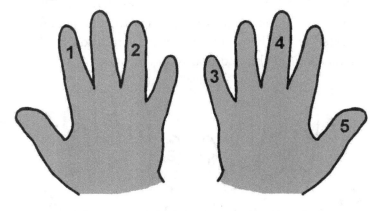

First pass with eight times table for 9×8.

Eight times table. Let us now calculate the result of 9×8. Recall that we must keep in mind the calculation position in the eight times table (9). With the eight times table, we jump two fingers at a time from the left, performing a second pass through the fingers to get to nine.

The finger that we reach in this way is dropped.

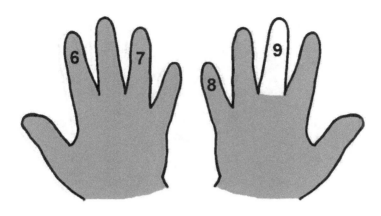

Second pass with eight times table for 9×8.

As before, the units digit in our two-digit result is the number of fingers that are still raised to the right of the finger dropped (2). The tens digit is calculated from the position in the table (9) minus the number of times we pass along the ten fingers and since we drop the finger on the second time round, this is 2. So the tens digit arises from $9 - 2 = 7$. This gives the result $9 \times 8 = 72$.

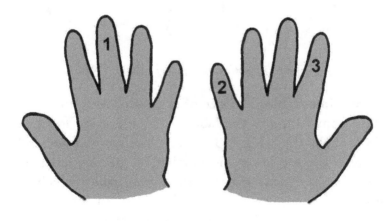

First pass with seven times table for 8×7.

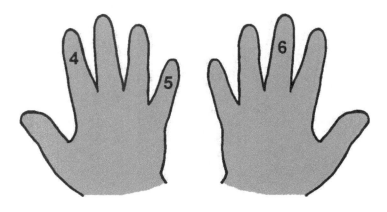

Second pass with seven times table for 8×7.

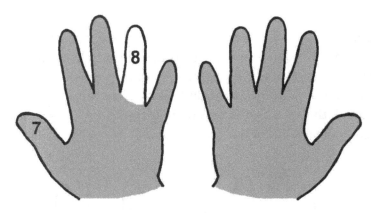

Third pass with seven times table for 8×7.

Seven times table. The size of jump used in counting results from 10 minus the table we are using. So for the 7 times table, the jump is 3. Clearly, the method demands more effort the lower the number of the multiplication table but is easier to use on the higher tables which are usually the most troublesome. For the seven times table, the 8×7 calculation is left as an exercise. However, can you explain why the method works?

Creative Thinking Puzzles I

Each puzzle has two hints at the back of the book.

11. Mad House

Every day the postman passes a certain house and wonders who might own it. The following is a list of attributes that could be ascribed to the house owner:

(a) sad (b) angry (c) relaxed (d) manic (e) confused

Which one is the most appropriate? [Solution 2]

12. Right Angle

EWXHUO

Most mathematicians will be unaware that the right angle suggests a certain six-letter country. Can you find it from the given letters? [Solution 30]

13. Amazing

Can you find a route through the maze from A to B?
[Solution 27]

14. The Dead Dog

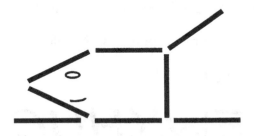

Shown is a dog lying down, facing to the left, with its tail
in the air. Can you rearrange one stick so that the dog looks
dead? [Solution 49]

15. Sum Line

$$9 - 8 + 4 = 6$$

Can you remove one straight line to make the equation correct? [Solution 8]

16. Which Way?

A cyclist comes to a crossroads and sees a sign. Can you add a single arrow to the middle circle to indicate the correct direction he should go? [Solution 4]

17. Nothing for It

Can you add one of the pair of letters below the equals sign to the left-hand side of the equation and the other of the pair to the right-hand side, then rearrange the four letters on each side into words to make an equality that states that "nothing equals nothing"? [Solution 32]

18. Rough Graph

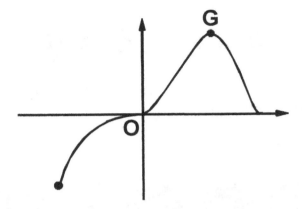

What large capital letter should be placed to the right of the lower left point? [Solution 17]

19. The Lighthouse

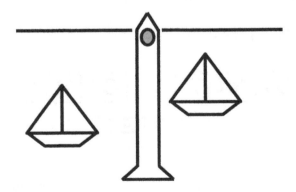

Shown are two identical boats, a lighthouse, and a horizon. Can you rearrange the two boats to show that they carry an identical cargo? [Solution 12]

20. The Pig and the Bird

Porky the pig, whose head is shown, was enjoying a nap when a bird accidentally flew into him. Can you add one triangle to the picture to show the bird battered and bruised? [Solution 45]

The Sleeping Beauty Problem

In 1997, Michele Piccione and Ariel Rubinstein[1] published a problem in *Games and Economic Behaviour* entitled "The Absent Minded Driver". It was intended to illustrate how beliefs could be determined in situations of imperfect recall and, without committing to a position, they outlined two different possible answers. Three years later, Adam Elga[2] discussed the problem in *Analysis* and, following Robert Stalnaker, named it The Sleeping Beauty Problem. Those who subscribe to a view on the issue usually fall into one of two groups, halfers and thirders, depending on the probability answer 1/2 or 1/3 that they arrive at. Elga formulated the problem as follows.

> Some researchers are going to put you to sleep. During the two days that your sleep will last, they will briefly wake you up either once or twice, depending on the toss of a fair coin (Heads: once; Tails: twice). After each waking, they will put you back

to sleep with a drug that makes you forget that waking. When you are first awakened, to what degree ought you to believe that the outcome of the coin toss is Heads?

Elga points out that the drug is so constituted as to return the subject's mind to its state before the coin toss and that the subject fully understands the conditions of the experiment. Sleeping Beauty (SB) is the identity of the unfortunate victim.

At first glance there appears to be a difficulty with the idea "when you are first awakened" because SB has no way of knowing which awakening is the first one. So we interpret this to mean "just after being awoken" and Elga appears to use it in this sense throughout. In fact, in the version sent to the newsgroup *rec.puzzles* in 1999, Jamie Dreier instead posed the question as "What is your credence now for the proposition that our coin landed Heads?" thereby removing this ambiguity.

Now there are two distinct possibilities to consider. The first relates to SB's estimate of the probability that the coin landed Heads when it was tossed, taken over all coin tosses; the second is connected with SB's chance of being correct on stating Heads at every awakening, taken over all awakenings. In other words, we either count the coin tosses or the awakenings.

Counting coin tosses. Once the coin has been tossed, the experimenters can assign a value of either 0 or 1 to the probability that the coin landed Heads. On awakening, SB has no more information as to what the outcome was than she had at the start of the experiment. If she is required to estimate the chance that the coin landed Heads she must say 1/2.

Counting awakenings. If the question is not about the coin landing but about the chance of SB being correct if she states Heads at every awakening, then the analysis is different. SB knows that she will be awoken twice when Tails occurs but only once when a Head falls. So if during the experiment she states Heads at every awakening then she knows that she will be correct in one guess out of three. The chance that SB will correctly state that the coin is Heads if she always utters Heads is now 1/3.

We note the difference in emphasis: the first case is about the fall of the coin regardless of what happens afterwards; the second is about the structure of the experiment and concerns the consequence of SB having one chance of being correct for Heads against the two permitted for Tails.

The next obstacle that must be overcome is to clarify which of these two possibilities the problem, as posed by Elga, is referring to. Both Dreier's question, "What is your credence [probability estimate] now for the proposition that our coin landed Heads?", and Elga's formulation, "to what degree ought you to believe that the outcome of the coin toss is Heads", are evidently about the event of the coin landing Heads. There is nothing that refers to SB's success rate in stating Heads on being awakened. So since the coin falls Heads in one out of two tosses, the answer to The Sleeping Beauty Problem is 1/2, a result with which Elga disagreed. Unfortunately, he also believed that he had found a counter-example to Bas van Fraasan's Reflection Principle which perfectly summarizes The Sleeping Beauty Problem:

Any agent who is certain that she will tomorrow have credence x in proposition R (though she will neither receive new information nor suffer any cognitive mishaps in the intervening time) ought now to have credence x in R.[3, 4]

In other words, if there is nothing to change SB's pre-experiment estimate of the probability that the coin landed Heads (which is 1/2) then it will not change during the experiment.

Our conclusion is that the questions posed by Elga and Dreier are not about the relative frequency with which SB will be correct if she states Heads at every awakening. They are about the relative frequency with which the coin will land Heads which is clearly 1/2.

References

1. Piccione, M. and Rubinstein, A. "On the interpretation of decision problems with imperfect recall", *Games and economic Behaviour*, **20**, (1969), 3-24.

2. Elga, A. "Self-locating belief and The Sleeping Beauty Problem", *Analysis*, **60**(2), (2000), 143-7.

3. van Fraasan, B. C. "Belief and the will", *Journal of Philosophy*, **81**, (1984), 235-256.

4. van Fraasan, B. C. "Belief and the problem of Ulysses and the Sirens", *Philosophical Studies*, **77**, (1995), 7-37.

Logic Puzzles

21. Shilly Chalet

At Jollywobble holiday camp, chalet numbers one, two and three were awaiting the six guests. Two men were to occupy each chalet, however, they had difficulty agreeing about the sharing arrangements. When Keith noticed that the toilet was missing, he rejected chalet number two next door. Then Neil turned down chalet number three next door to number two on the grounds that it smelt like a cesspit. Eventually, Gordon shared with neither Martin nor Ian. Ian shared with neither Keith nor Colin. Also, Neil found that he was next door to neither Gordon nor Ian. Which pairs occupied which chalets? [Solution 36]

22. The Five Chimneys

In the Beijing district of Foo Mee Gate, there is a red-bricked terrace where the residents burn coal fires. High up on the roof of number 692, five chimney pots A, B, C, D, E sit in close proximity to each other. It turns out that each chimney is capable of drawing in (−) or sending out (+) 1, 2, 3, 4, 5, units of smoke each day, respectively. The local chimney cleaning company, Soo Tian Sweep, noted that each chimney controls the behaviour of only one of two other chimneys where the relationships are as follows. Chimney −5 gives −1 or −4, that is, if chimney E draws in 5 units then either chimney A takes in 1 unit or chimney D takes in 4 units. Also, −4 gives −1 or +5; −3 gives +2 or +5; −2 gives −3 or −4; −1 gives +3 or +4; +1 produces −2 or −5; +2 gives −5 or +1; +3 gives +1 or +4; +4 gives −3 or −2; and +5 gives +2 or +3. To remain habitable, house number

692 must send out more smoke than it draws in. What is the direction of smoke (+ or −) for each chimney? [Solution 10]

23. A Pressing Problem

	1	2	3	4
A	red	yellow	blue	green
B	red	green	yellow	blue
C	blue	red	green	green
D	green	yellow	blue	red

The Department of Silly Ideas has 16 push buttons on the front door arranged in a 4×4 square as shown. Each button is lit in one of the four colours in the rotating sequence: blue, red, yellow, green. When a button is pressed, its colour moves one forward in the sequence (e.g. pushing red makes it yellow, pushing green makes it blue etc.). In addition, three pairs of buttons each have an internal relationship where one button controls the colour of the other one but not vice versa. When the controlling button of the pair is pushed, its colour moves one forward while the controlled button colour moves one backwards. In the panel shown above, B1 controls C3, A3 controls D1, and D2 controls B4. So, for example, pressing A3 sends A3 from blue to red and D1 from green to yellow. However, pressing D1 just sends D1 from green to blue. Now entry into the department is gained by producing the same colour for all the buttons. What is the smallest number of button pushes required to gain entry? [Solution 26]

53

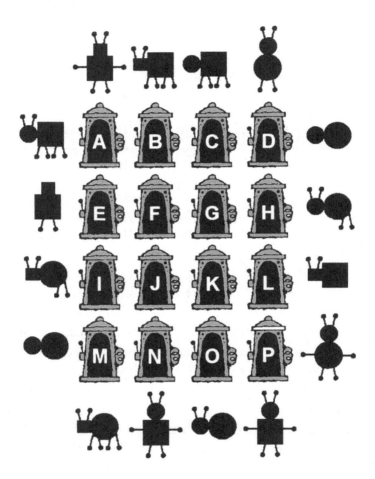

Shown are 16 mutation chambers labelled A–P, surrounded by alien figures. Each of the four aliens on the left has passed through the four chambers directly to their right and has been transformed into the alien on the far right (e.g. the figure to the left of A has moved through chambers A, B, C and D to finish as the one to the right of D.) Similarly, each of the four aliens shown above the chambers has passed through the four chambers directly below to finish as seen at the bottom. It's known that each chamber always effects a single alteration (e.g. changes head or body shape, changes posture, adds or removes appendages). Neither a row nor a column contains more than one of any type of chamber. What does each chamber do? [Solution 11]

25. Santa Flaws

	Dwarf	Reindeer	Gift	House
1	Doc	Dancer	TV	Whywurry
2	Grumpy	Prancer	Doll	Dungroovin
3	Happy	Vixen	Radio	Binsleepin
4	Sleepy	Comet	Computer	Nomunie
5	Bashful	Cupid	Guitar	Takerhike
6	Sneezy	Donner	MP3 Player	Litesout
7	Dopey	Blitzen	Book	Ronguns

Santa has lent each of the seven dwarfs one of his reindeer to deliver a gift to a house in Hollyville. Unfortunately, the order the dwarfs should leave the grotto together with their delivery instructions has been mixed up in all the excitement. Although each item is in the correct column, only one entry in each column is correctly positioned. The following facts are true about the correct order.

(1) Happy is somewhere above Vixen

(2) Dungroovin is not above the TV

(3) The MP3 player is one above Litesout which is two above Takerhike

(4) Blitzen is two below Bashful and two above Whywurry

(5) Sleepy is three above Comet which is two below the guitar and three below Nomunie

(6) Donner is one below the radio and one above Dopey

(7) The doll is one below Dancer and two above Sneezy

Can you find the correct dwarf, reindeer, gift, and house name for each position? [Solution 7]

26. Safe Cracker

Four different digits 1–9 are required to open a safe. The first and third numbers total ten; the second number is no more than four; the third number is a prime number; the fourth number is three less than the first; and the second number is one less than the fourth. Can you find the four-digit safe combination? [Solution 6]

27. A Raft of Changes

Twelve special agents, all chosen to have an equal weight, were sitting on a raft ready to embark on a secret mission. In order to balance the raft in the water the men were divided into three groups: ABCD sat at the back, EFGH in the middle, and IJKL at the front. However, just after setting off, the group leader looked over the side and noticed that the raft tilted forwards. He failed to observe that the inclination was that which might be expected from two men having a different weight from the rest. Assuming that there was only one spy on the raft, he surreptitiously consulted his manual and discovered that for this situation two tests could be carried out which would always identify the imposter and would also show whether he was heavier or lighter than the chosen weight.

For the first test only, four interchanges were made involving ABCEFGIL so that the group leader was situated in a group that was not adjacent to one containing either A or H. Noting the inclination of the raft, the leader then rearranged the team for the second test in three interchanges, without changing his own position, so that CFHI finished at the back of the raft, DEKL in the middle, and ABGJ at the front. From the resulting inclination he managed to deduce who the spy was and to his great surprise found it was himself! Mistrusting his own logic, he decided to continue with the mission, and on reaching the destination, he subsequently caught two of the team trying to send a radio message to the enemy in a nearby wood. One was heavier than the chosen weight by as much as the other was lighter. Who were the two enemy agents? (See following notes.)

Notes

- A "group" is defined as four men sitting in a straight line perpendicular to the direction of travel of the raft.

- An interchange involves the juxtaposition of two men from different groups in the raft.

[Solution 29]

Parity Tricks with Coins

Trick 1

This effect requires an assistant and six identical coins.

Spread out the six coins on the table and ask your assistant to randomly turn over any number of coins. You then turn your back on him and provide him with the following instructions.

(1) He must select any digit from 1–6 inclusive and tell you what it is.

(2) He must then make that number of coin turns on the six coins, where a turn changes a Head to a Tail or vice versa. The same coin can be turned more than once if desired which of course then counts as more than one turn.

(3) The positions of the coins on the table can be freely mixed up though they must be kept separate.

(4) Finally, he must cover one of the coins with his hand.

When you turn around you are able to announce whether his hidden coin is a Head or a Tail. Unfortunately, your assistant is not impressed because your chance of being correct is one half. A lucky guess perhaps? Not so when you show you are able to repeat the feat again and again, each time giving the correct answer. The chance of being correct diminishes as $1/2^n$, where n is the number of attempts, so it won't take long to impress the audience!

Secret

The trick works on the principle of parity. Just before you turn around, secretly count the number of Heads showing and note if this number is even or odd (zero is taken to be even). We shall call this the "Heads parity". Note also whether or not the number of turns that your assistant informs you he will make is even or odd. If this last number is even then the Heads parity will be unaltered, in other words, if the number of Heads was even when you last looked then you can expect there to be an even number of Heads when you turn around. Similarly for an odd number of Heads. However, if the number of turns is odd the Heads parity will change from even to odd and vice versa. When you turn around simply note the number of visible Heads and decide from parity considerations whether or not there should be another one beneath his hand.

Trick 2

This effect requires an assistant and any six coins.

With your back to the table, ask your assistant to select any six coins and lay them out on the table in a line Tails up. For the purpose of the trick, he must mentally number them from 1–6 from left to right. Let him turn one of these into a Head and note its position, then ask him to think of any number greater than 6 which will then be the number of switches he is about to make. Without telling you these two numbers, he must add them and tell you the total. So if the Head is at position 5 and he chooses to make 17 switches then the only number he should tell you is their total 22.

A switch consists of juxtaposing the Head with any adjacent coin and your assistant must now perform his chosen number of switches. When he has finished, without looking, you are able to tell him which coins to discard so that only the Head is left on the table.

Secret

Even number of coins. If the total of position and switches given by the assistant is even, ask him to remove the coin at the left end of the row; if it is odd he must eliminate the coin at the right end. The instructions to your assistant that follow are then the same for both cases.

(1) Switch the Head with an adjacent coin.
(2) Remove the two coins at the ends of the row.

Instructions (1) and (2) are then repeated and the Head will be the only coin left on the table.

This procedure works with any number of even coins whether you are told the number of them or not. As above, first eliminate an end-of-row coin according to whether the given total is even or odd, then simply keep repeating (1) and (2) until only the Head remains. If there are n coins (n is even) then the number of applications of (2) is $(n-2)/2$ although this fact will be unknown to the conjuror if he does not request knowledge of the number of coins.

Odd number of coins. This effect also works with any number of odd coins whether or not you know their number. This time, if the total of position and switches is even, first the two end coins are removed; if it is odd, no initial action is taken. Instructions (1) and (2) are then carried out until the coins are reduced to a single Head. Either way, if there are n coins (n is odd) then (2) is applied $(n-1)/2$ times.

Of course, if the number of coins chosen by the assistant is not requested by the conjuror, he must at least obtain the fact that their number is even or odd to decide whether the "even number" or "odd number" procedure should be adopted. This information could be given by a confederate who might tap his foot to signal an even number. Also the number of applications of (2) will be unknown and the conjuror must continue the repeated application of (1) and (2) until the assistant or a confederate states that there is only a Head left on the table.

The first trick is a more complex variation of one that appears in Walter Gibson's *Professional Magic for Amateurs*[1] while the second is an enhancement of Jack Yates's trick with four matches which appears in his book *Minds in Close-up*.[2]

References

(1) Gibson, W., *Professional Magic for Amateurs* (Dover: 1974).

(2) Yates, Jack, *Minds in Close-up* (Supreme Magic Co., USA: 1969).

The Shakespeare Puzzles

There is evidence that Shakespeare was partial to inserting topical references in his work and used a simple puzzle in the process. In the Introduction to the 1923 Cambridge edition of *Love's Labour's Lost*, an argument is given for the character Moth representing the English pamphleteer Thomas Nash.

> *Armado.* How canst thou part sadness and melancholy, my tender Juvenal?
> *Moth.* By a familiar demonstration of my working, my tough signior.
> *Armado.* Why tough signior? Why tough signior?
> *Moth.* Why tender Juvenal? Why tender Juvenal?
> (1594 *Love's Labour's Lost*, 1.2.7–8)

In both Robert Greene's *A Groatsworth of Witte* (1592) and Frances Meres's *Wit's Treasurie* (1596), Thomas Nash is likened to the Roman poet Juvenal with "sweet Tom" and "Young (juvenile) Juvenal". In 5.1.63–4, of *Love's Labour's Lost*, Costard calls Moth "thou halfpenny purse of wit, thou pigeon-egg of discretion", a reference to the Nash–Harvey controversy when, in *Pierce's Supererogation* (1593), Gabriel Harvey labels Thomas Nash "a young man of the greenest springe, as beardless in judgement as in face, and as Peniless in wit as in purse" with the suggestion that he might next "publish Nashe's [sic] *Penniworth of Discretion*". Shakespeare's puzzle is simply to make Moth an anagram of Thom!

There is a growing trend in these times to doubt that William Shakspeare of Stratford wrote the work attributed to him. The argument is that an ambitious courtier with designs on high office used him as a mask so that his prospects of promotion would not be impeded by being exposed as a dramatist, a working-class profession in those times. It is supposed that the real author left his signature to posterity somewhere in the Shakespeare work. One candidate often suggested is Sir Francis Bacon, a leading Renaissance philosopher, who rose to become Lord Chancellor and to Ben Jonson was "by his work, one of the greatest men ... that had been in many ages".[1] There are some interesting occurrences of Bacon's name in the dedications that preface the Shakespeare collections. We shall examine three dedications: one from *Shake-speare's Sonnets* (1609), and two from the First Folio (1623) collection of 36 plays. Whether or not these messages were intentionally placed there is left for the reader to decide.

Sonnets Dedication

The *Shake-speare's Sonnets* dedication has even been recognised by academics as a puzzle and it invites interpretation (see Figure). Most of their attention has been focused on the identity of M^r. W.H., however, the approach here will be to look for a simple acrostic device. An acrostic uses the first or last letters of every word or line to spell out a concealed message, although strictly speaking, if the last letters of words are used instead of the first, the device is called a 'telestich'.

The first feature of this puzzle that merits attention is that the points between words resemble those of a Roman inscription. The original Latin alphabet had 21 letters as follows:

A B C D E F Z H I K L M N O P Q R S T V X

Around 250BC the Z was replaced with a G to leave:

A B C D E F G H I K L M N O P Q R S T V X

TO.THE.ONLIE.BEGETTER.OF.
THESE.INSVING.SONNETS.
Mr.W.H. ALL.HAPPINESSE.
AND.THAT.ETERNITIE.
PROMISED.

BY.

OVR.EVER-LIVING.POET.

WISHETH.

THE.WELL-WISHING.
ADVENTVRER.IN.
SETTING.
FORTH.

T. T.

Substitution cipher acrostic puzzle purportedly containing Bacon's name and his current business, T.T. dedication, Shake-speare's Sonnets (1609)

This would have been the 21-letter alphabet used by Julius Caesar (c.100–44BC) for his Caesar Cipher when he sent encoded messages to his battlefield generals, most notably

Cicero. Suetonius, in his *Lifes of the Caesars LVI* from the 2nd century AD, describes Julius Caesar's simple cipher:

> There are also letters of his to Cicero, as well as to his intimates on private affairs, and in the latter, if he had anything confidential to say, he wrote it in cipher, that is, by so changing the order of the letters of the alphabet, that not a word could be made out. If anyone wishes to decipher these, and get at their meaning, he must substitute the fourth letter of the alphabet, namely D, for A, and so with the others.[2]

In other words, the letters in the coded message are shifted +3 in the alphabet to obtain the real message. Plutarch's *Life of Julius Caesar* supports the account given by Suetonius:

> And it is thought that he was the first who contrived means for communicating with friends by cipher, when either press of business, or the large extent of the city, left him no time for a personal conference about matters that required dispatch.

It is possible that Francis Bacon knew of this practice for in his *Character of Julius Caesar* he informs us that:

> For his own person he had a due respect: as one that would sit in his tent during great battles and manage everything by messages.[3]

After the invasion of Greece in the first century BC the letters Y and Z were added, and in Elizabethan times J, U, and W were introduced. The Tudors used I and J as well as U and V interchangeably, while W was interpreted as being two U or two V. Penn Leary, a trial lawyer from Omaha, produced the name BACON in the *Sonnets* dedication from the following words:

OF.THESE.INSUING.SONNETS.M[r].

Leary selected the last letter of each word FEGSR and took the word 'FORTH' (fourth) in the dedication as a shift indicator thereby displacing each letter four places backwards to reveal BACON. Unfortunately, Leary could make no further progress with the puzzle.

It is suggested here that the solution falls into three parts and that the first part, which includes the first four words, are to be taken at face value. So the first two parts of the solution read:

TO THE ONLIE BEGETTER BACON …

where BEGETTER means 'originator'. For the third part of the solution, consistent with Leary's method, we suggest taking the final letter of each of the remaining 23 entries but this time without applying a shift. Here, an entry consists of a letter or string of letters bounded at each end by points, noting that the hyphens in 'EVER-LIVING' and "WELL-WISH-ING' resemble points. So, for example, the entries EVER and LIVING are considered to be separate and contribute R and G. Starting with .W., this yields the letter string:

WHLEDTEDYRRGTHELGRNGHTT

The printer was Thomas Thorpe and the last two T appear to represent his name but the letters are bounded by points so we allow their contribution to the solution. We now partition this string as follows:

WHLE/DTED/YR/RG/THE/LGR/NGHTT

and take this as an invitation to insert vowels for sense. A fair attempt would be:

According to *Webster's Revised Unabridged Dictionary* (1913), the now-obsolete LEGER (also 'leiger' or 'lieger') was "a minister or ambassador resident at a court or seat of government". For example, in modern times, the US Leger to the UK would live in London where the government of the UK resides. The term appears in Shake-speare's *Measure for Measure*:

> *Isa.* ... Lord Angelo hauing affaires to heauen
> Intends you for his swift Ambassador,
> Where you shall be an everlasting Leiger;
> (1604 *Measure for Measure*, 3.1)

as well as in Sir Francis Bacon's letter to his friend Toby Matthew in the summer of 1609 "on the other side it is written to me from the leiger at Paris".[4]

The term 'ADVENTURER' in the *Sonnets* dedication refers to one who took on a shareholding risk and it appears in the Second Virginia Charter of 23 May 1609, a document that lists the shareholders and governing members of the new Virginia Colony:

> and incorporated by the Name of The Treasurer and Company of Adventurers and Planters of the City of London, for the first Colony in Virginia.

The book of *Shake-speare's Sonnets* was recorded in the Stationers Register on 20 May 1609, thereby declaring the intention to publish (and works were usually printed within a year of entry). These two events, the *Sonnets* registration and the publication of the Second Virginia Charter, occurred three days apart so the most significant adventuring at the time of registering the *Sonnets* was investment in the Virginia colony. Sir Francis Bacon, who was already advising King James on plantations in Virginia, was named on the charter as one of about 50 Council members charged with governing the colony (and as Solicitor General, he must have been a prime mover). The most interesting point, however, is that the government resided not in Virginia but in London:

> Therefore we Do ORDAIN, establish and confirm, that there shall be perpetually one COUNCIL here resident, according to the Tenour of our former Letters-Patents; Which Council shall have a Seal for the better Government and Administration of the said Plantation, besides the legal Seal of the Company or Corporation, as in our former Letters-Patents is also expressed.

Since Sir Francis Bacon was an ambassador for the Virginia Colony and commuted to London, the seat of the Virginia government, this would have made him a leger knight. So, taking REG to mean REGISTERS, the third part of our message becomes:

WHILE DATED YEAR REGISTERS
THE LEGER KNIGHT

Our interpretation shall be that the year (1609) together with the date (20 May) of registration of the *Sonnets* virtually coincides with the occasion when the knight, Sir Francis Bacon, became a Leger. So we claim here that the complete message runs as follows:

TO THE ONLIE BEGETTER BACON WHILE DATED YEAR REGISTERS THE LEGER KNIGHT

First Folio I.M. Dedication

One of the dedications at the front of the First Folio 'To the memorie of M. *W. Shake-speare.*' is signed I.M. which some commentators have speculated to be James Mabbe (see Figure).

The verse itself appears cryptic and is reminiscent of an observation by Bacon in his *Advancement of Learning*:

> As we see in Augustus Caesar, (who was rather diverse from his uncle, than inferior in virtue) how when he died, he desired his friends about him to give him a *plaudite* [italics added], as if he were conscient to himself that he had played his part well upon the stage.[5]

Geometrical substitution cipher puzzle, 'To the memorie of M. W. Shake-speare', dedication by I.M., Shakespeare's Comedies, Histories & Tragedies (1623)

The piece has two characteristics in common with the *Sonnets* dedication: it is signed by initials instead of a full name; and one line of the verse contains the word 'forth' which we claimed was a shift indicator in the *Sonnets* solution.

Apart from the last, the lines decrease in length and if it *is* a puzzle this almost suggests a geometrical solution. In fact, close inspection reveals that the capitals FWSG in the second line of the verse, S in the fourth line, E in the seventh line, and R in the bottom line can be connected by a triangle [added in Figure] with the oblique line almost angled in sympathy with the right-hand side of the dedication. This creates the Roman set WSFEGSR (the A in "An" in line 5, which is almost in line, anyway appears as a different font). The first two letters WS might be interpreted as William Shake-speare while FEGSR taken with the word "forth" as a shift indicator, as in the *Sonnets* solution, produce the name BACON. If one "wen'st but forth" in applying the cipher shift then it is BACON who will "enter with applause" and receive the "Plaudite".

Jonson's First Dedication

Ben Jonson has two tributes in the First Folio, the second of which has already been discussed. His first tribute sits on the left page opposite a wood-cut image of Shake-speare that is traditionally taken to be an accurate likeness.

The tribute shares an idea that also appears on a portrait miniature of Francis Bacon painted in Paris by Nicholas Hilliard, England's leading miniaturist, for Bacon's eighteenth birthday. The Latin inscription Hilliard inserted around the face reads *si tabula daretur digna animum mallem*, that is, "if the face as painted is deemed worthy, yet I prefer the mind."

The second observation is that it also shares a feature of the *Shake-speare's Sonnets* and I.M. puzzles in that the signature at the foot of the piece appears in initials, in this case B.I. One might expect that the very first dedication would identify its author explicitly so the choice of initials is curious. These reasons raise the suspicion that we are dealing with a concealment cipher and the unwarranted punctuation after "Nature", "O", "brasse" and "All" hints at the use of punctuation indicators.

To the Reader.

This Figure, that thou here feeſt put,
  It vvas for gentle Shakeſpeare cut;
VVherein the Grauer had a ſtrife
  with Nature, to out-doo the life :
O, could he but haue drawne his wit
  As well in braſſe, as he hath hit
His face ; the Print would then ſurpaſſe
  All, that vvas euer vvrit in braſſe.
But, ſince he cannot, Reader, looke
  Not on his Picture, but his Booke.

B. I.

Punctuation puzzle containing Bacon's name, dedication by Ben Jonson, Shakespeare's Comedies, Histories & Tragedies (1623)

One example of this type of concealment dates from Cromwell's time, less than 20 years after the publication of the First Folio. Sir John Trevanion was imprisoned in Colchester Castle ready to meet his execution for extending his sympathy to the Royalists. Despite being under constant guard and his correspondence closely scrutinised, his friends still managed to get a message through to him. The message he received ran as follows:

> Worthie Sir John:- Hope, that is ye beste comfort of ye afflicted, cannot much, I fear me, help you now. That I would saye to you, is this only: if ever I may be able to requite that I do owe you, stand upon asking me: 'Tis not much I can do: but what I can do, bee you verie sure I wille. I knowe that, if dethe comes, if ordinary men fear it, it frights not you, accounting it for a high honour, to have such a rewarde of your loyalty. Pray yet that you may be spared this soe bitter, cup. I fear not that you will grudge any sufferings; onlie if bie submission you can turn them away, 'tis the part of a wise man. Tell me, as if you can, to do for you anythinge that you can wolde have done. The general goes back on Wednesday. Restinge your servant to command.   R.T.[6]

The message seems perfectly innocent until one takes the third character after each punctuation mark to reveal:

Panel at east end of chapel slides

That evening, while alone at prayer in the chapel, Sir John made his escape.

In *Elementary Cryptanalysis: A Study of Ciphers and Their Solutions* we learn that the use of punctuation marks to conceal a message is a known practice:

> Significant letters may be concealed in an infinite variety of ways. The key, as here, may be their positions in words, or in the text as a whole. It may be their distance from one another, expressed in letters or in inches, or their distance to the left or right of certain other letters (indicators) or of punctuation marks (indicators); and this distance need not be constant or regular.[7]

Returning to the Shake-speare tribute, perhaps the "Figure" referred to is not the Shake-speare face on the opposite page but the comma that immediately follows the words "Figure" and "put":

This Figure, that thou here seest put,

thereby drawing attention to the commas in the text. If we now select the first letter of each word that ends with a comma, the following letters arise: FpNObABcRP which can rearrange as FRbAcONpPB to give

FRbAcON [Francis Bacon] pP [*per procurationem*, by delegation to] B [Benjamin]

This appears more credible considering it is Ben Jonson's tribute and the use of B to represent 'Ben' is justified given that his name appears as B.I. This would then suggest that Ben Jonson was employed by Sir Francis Bacon to oversee the production of the Shake-speare First Folio.

Of course, interpretations such as these cannot be used as evidence of alternative authorship. If one grants that they are real puzzles — and I suspect that they are — how can one be sure that from the many possible enciphering methods that human ingenuity might invent, the ones used here were the intended ones? It is this freedom of selection that renders all Shakespeare authorship decipherments questionable, because the charge can always be made that the liberty of method allows the contrivance of whatever solution one desires. This is not to say that the Shakespeare authorship question should not be investigated, only that the standard of evidence should be high enough so as to be admissible in a court of law.

References

1. Jonson, Ben, *Timber: or Discoveries* (1641), Workes, Bodleian Library, Oxford, (Douce.I.303.STC 14753).

2. Kahn, David, *The Codebreakers: The Story of Secret Writing* (New York: Scribner, 1996), p.77.

3. Spedding, James, *Works of Lord Bacon*, Vol. 6., p.344 (1857-9).

4. A Collection of Letters by Sr Toby Matthew Knight, printed for Sir Henry Herringham 1660, p.13; York Minster Library, UK, MS XXXV.G.36.

5. Bacon, Francis, Negotiation in Civil Philosophy (1605), *Francis Bacon The Major Works* (Oxford University Press: 2002), p.272.

6. Deacon, R., *A History of the British Secret Service* (Frederick Muller: 1969), pp.43-4.

7. Fouché Gaines, Helen, *Elementary Cryptanalysis : A Study of Ciphers and Their Solutions* (Chapman and Hall: 1940).

Creative Thinking Puzzles II

Each puzzle has two hints which can be found at the back of the book.

28. Inspector Lewis

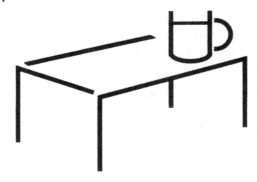

At the scene of the crime, PC Karel found the Inspector looking at a glass table with a mug on it. Who did they belong to? [Solution 56]

29. The Builder's Problem

On the left is a builder's hod full of sand next to an empty box. The sand in the hod is poured into the box. Can you rearrange two straight lines to show the box half full and the entire hod higher than the box base? [Solution 22]

30. The Four Dice

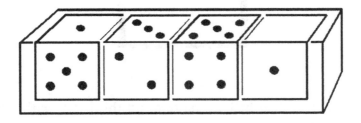

What is the missing number of spots on the blank face? [Solution 39]

31. Politically Correct

$$| + O \geqslant |O$$

How can the pieces be rearranged to make a politically correct equality? [Solution 15]

32. Inklined

It takes less than half a minute to fill this bottle with ink and less than a minute to evacuate it. By tipping the bottle can you discover how long it takes to empty? [Solution 57]

33. Bubble Math

Di is blowing bubbles using a pipe. Can you relocate exactly one bubble to correct the diagram? [Solution 20]

34. Arithmystic

$$1 + 98 - 10 - 89 = 0$$
$$1 + 98 = 10 + 89$$

The mathematics teacher Mark Wright wrote the above two equations on the white board then announced that only the first was correct. What was his reasoning? [Solution 51]

35. No Escape

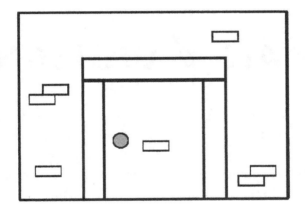

Shown is a small roofed prison cell with no windows and a closed door at the front. Can you relocate the door knob to reveal why the prisoner can never leave his cell? [Solution 42]

36. Secret City

I LOVE
MY CITY

The graffiti shown was written on a wall somewhere in the USA. Can you delete one straight line to reveal the location of the message? [Solution 54]

37. Winning Line

MAN **V** WOMAN

A husband and wife have an argument which the wife wins. Can you remove four letters, at least one from each side of the V, and draw one straight line to underline the fact that only the woman remains, victorious? [Solution 13]

The Hardest Ever Logic Puzzle

Although the following is a new puzzle, its method of solution can be derived from a similar puzzle credited to George Boolos which is often referred to as "The Hardest Logic Puzzle Ever".

> A prisoner in solitary confinement, who has no idea what day of the week it is, has been told that on each day for three consecutive days he can ask the guard on duty a question to which he will receive a 'yes' or 'no' answer. From the responses, he must announce on the third day what day of the week it is. If he is correct he will be set free, if not he will be executed.
>
> The prisoner is faced with several difficulties. A guard will either consistently tell the truth or consistently lie and on no two days of a week will the guard be the same man. Furthermore, although the guards understand English, they have been instructed to give a 'ho' or a 'mo' answer, one of them meaning 'yes' and the other meaning 'no'. The prisoner has no idea which is which.
>
> What line of questioning is sure to win the prisoner his freedom?

In 1992, a book by Boolos entitled *La Repubblica* presented a logic puzzle under the heading *L'indovinello più difficile del mondo* ("The most difficult enigma in the world"). It was evidently inspired by Raymond Smullyan's *What is the name of this Book?*[1] where problems appear involving truth-tellers and liars who cannot be discriminated. To complicate matters, the "yes" and "no" answers that are given to questions asked of the truth-tellers and liars are uttered in their own language and cannot be understood. Boolos gave his problem as follows:[2]

Three gods A, B, and C are called [named], in some order, True, False, and Random. True always speaks truly, False always speaks falsely, but whether Random speaks truly or falsely is a completely random matter. Your task is to determine the identities of A, B, and C by asking three yes-no questions; each question must be put to exactly one god. The gods understand English, but will answer all questions in their own language, in which words for 'yes' and 'no' are 'da' and 'ja' in some order. You do not know which word means which.

The following clarifications were also provided:

(1) It could be that some god gets asked more than one question (and hence that some god is not asked any question at all).

(2) What the second question is, and to which god it is put, may depend on the answer to the first question (and of course similarly for the third question.)

(3) Whether Random speaks truly or not should be thought of as depending on the flip of a coin hidden in his brain: if the coin comes down heads, he speaks truly; if tails, falsely.

(4) Random will answer 'da' or 'ja' when asked any yes-no question.

The key to solving this puzzle lies in realizing that Random does not simply utter a "da" or "ja" at random without considering the question. He is actually taking on the role of either a truth-teller or a liar and is therefore fully processing the question asked according to his randomly adopted state. There is also a particular format one can use to frame a question in which the answer 'ja' can be associated with the affirmation of a statement X and 'da' with its negation. This format allows for the variation in state of Random and is as follows:

> In your present state, are you the kind of god who would say that only one of the following two statements is true: "X" and "da means yes"?

Let us apply this technique to the puzzle in view. The first information one can establish is whether or not A is Random. We ask A the following question:

> In your present state, are you the kind of god who would say that only one of the following two statements is true: "You are Random" and "da means yes"?

There are four cases to consider: A is True, A is False, A is Random speaking the truth, and A is Random speaking falsely. We tabulate the responses for the cases where 'da' means 'yes' and then 'ja' means 'yes'.

	da	**ja**
T	da	da
F	da	da
R_T	ja	ja
R_F	ja	ja

The top row gives the possible 'yes' word and the left column the possible identities for A. The responses demonstrate that one need not know which word means 'yes' because a 'da' means that A is a truth-teller or liar and a 'ja' means that A is Random.

Once A is known to be Random, only one further question is needed to identify all three gods. We ask B the following question:

Are you the kind of god who would say that only one of the following two statements is true: "You are True" and "da means yes"?

	da	**ja**
T	ja	ja
F	da	da

Of course, B can either be True or False and we tabulate the responses for each possible 'yes' word. A 'ja' is associated with B as True and a 'da' with B as False. This completely determines the three gods.

However, we must still consider the case where A is known to be either True or False. Here a further two questions are required, the first of which is directed to A as an enquiry about whether or not B is Random.

Are you the kind of god who would say that only one of the following two statements is true: "B is Random" and "da means yes"?

We now tabulate the four possible responses of A to this question relating to the identities of A, B, and C. Again the top row shows the possible words for 'yes'.

		da	**ja**
A	**T**	da	da
B	**F**		
C	**R**		

		da	ja
A	T	ja	ja
B	R		
C	F		

		da	ja
A	F	da	da
B	T		
C	R		

		da	ja
A	F	ja	ja
B	R		
C	T		

Here a 'da' means that B is True or False and C is Random while a 'ja' means that B is Random and C is True or False. By establishing whether A is True or False, a complete determination can be effected. To this end, we ask A a final question:

Are you the kind of god who would say that only one of the following two statements is true: "A is True" and "da means yes"?

	da	**ja**
T	ja	ja
F	da	da

God A can be either True or False.

Since a 'ja' means A is True and a 'da' means A is False then all three gods can be identified. We do not claim here that this is the only solution to the puzzle but it is certainly sufficient.

Returning to our day-of-the-week puzzle, we have now seen that it is possible to find out whether or not a statement X is true irrespective of whether a truth-teller or liar is responding and without knowing which word means 'yes'. All that is needed is a technique for narrowing down the possible days by asking questions about decreasing parts of the week. A good opening question on Day 1 would be as follows:

> Are you the kind of guard who would say that only one of the following two statements is true: "today is one of Sunday to Wednesday inclusive" and "ho means yes"?

We can construct a table for the guard being a truth-teller (T) or a liar (F) and for the possibilities of 'ho' being 'yes' and then 'mo' meaning 'yes'. If the day appears in "Sunday to Wednesday inclusive" then the results are as follows:

	ho	**mo**
T	mo	mo
F	mo	mo

If the day is in Thursday to Saturday inclusive the table becomes:

	ho	**mo**
T	ho	ho
F	ho	ho

A 'mo' response is associated with Sunday to Wednesday inclusive and a 'ho' is the other part of the week Thursday to Saturday inclusive. So if the response is 'mo' our new question on Day 2 could be:

> Are you the kind of guard who would say that yesterday, only one of the following two statements was true: "today is in Sunday or Monday" and "ho means yes".

This will establish by association whether Day 1 is one of Sunday or Monday, or one of Tuesday or Wednesday. The final question on Day 3 then enquires about the situation two days ago and the identity of Day 1 and hence Day 3 can be determined. Of course, if the first question resulted in a Thursday to Saturday inclusive result, then with luck it might be possible to establish the day with only one further question by choosing a two-day and one-day partition.

References

(1) Smullyan, Raymond. *What is the name of this Book?* Prentice-Hall: 1978.

(2) Boolos, George. "The Hardest Logic Puzzle Ever", *Harvard Review of Philosophy*, 6:(1996), 62-65.

Zeno and Infinitesimals

Certain aspects of space and time, even in the world of Aristotle (384–322BC), occupied thinkers to such an extent as to give rise to the invention of some interesting paradoxes. Problems such as whether or not a finite time period can contain an infinite set of points are still far from being settled in our own age. Connected with this issue is the dilemma as to whether or not the notion of the infinitesimal is valid, one which the English philosopher Bertrand Russell (1873-1970) defined as follows:

> a number or magnitude which, though not zero, is less than any finite number or magnitude.[1]

The problem with this definition can easily be seen if we call this special non-zero number N. If it is less than "any finite number" then that includes itself, so we have $N < N$ which is nonsense!

Around 450BC, the Italian Zeno of Elea constructed a set of problems in defence of his teacher Parmenides (c.475BC) who had maintained that change is impossible. Four of these problems are discussed by Aristotle in his *Physics* who attempts to counter them by a rigorous analysis of points (or "nows"), intervals, and infinity.

One of Zeno's paradoxes, usually referred to as "Achilles and the Tortoise", involves a tortoise which has been given d meters start in a race with Homer's mythological warrior Achilles. Both are moving at constant speed with the hero moving faster than the tortoise. By the time Achilles has covered the distance d, the tortoise has moved on by a lesser amount. Of course, once Achilles has negotiated this further distance, the tortoise has moved on again, and so on. Aristotle describes the problem as follows:

> the slowest runner will never be caught by the fastest runner, because the one behind has first to reach the point from which the one in front started, and so the slower one is bound always to be in front.[2]

The point of this problem has often been mistaken for suggesting than an infinite time is required for Achilles to catch the tortoise. Those who fall into this error usually calculate the time taken, find it to be finite, and so believe that the paradox is solved. However, by considering decreasing distances at constant speed Aristotle is adding decreasing times, and so is bringing time to a standstill in an 'infinity' of steps, whereas if we keep the steps in time constant then Achilles will succeed in overtaking the tortoise. Nevertheless, this is not the core of the issue. The actual point of the paradox is to argue that since an infinite number of catching-up processes is required, and one can never complete an infinity of tasks, then Achilles can never reach and overtake the tortoise.

This concept of 'infinitely divisible' also lies at the heart of the differential calculus where a curve is defined over a continuous interval. This interval is divided into a finite number of smaller intervals Δx in order to specify the gradient (slope) of a straight line (chord) between two points A and B on the curve. As an example, consider the curve $y = x^2$. By avoiding the assignment $\Delta x = 0$ (because a division by zero would be undefined), then by using high-school algebra we can set up the gradient equation for the chord AB on the curve as follows.

$$\frac{\Delta y}{\Delta x} = \frac{(x + \Delta x)^2 - x^2}{\Delta x} = 2x + \Delta x \qquad (1)$$

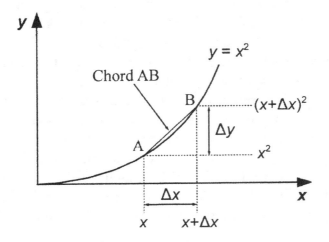

Gradient of chord AB.

This is not yet the differential calculus. This is the algebraic method of calculating the gradient of any chord AB on the curve $y = x^2$ given the x coordinate of A and the change in distance Δx required to reach the x coordinate of B.

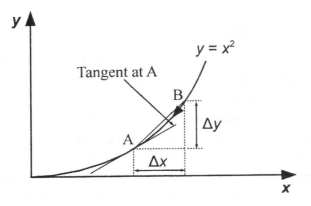

Tangent at A

The differential calculus arises when after fixing point A we decide to run point B down the curve towards it thereby making Δx and Δy decrease. The gradient of the chord AB then approaches the gradient of the tangent to the curve at A, the tangent being in contact with only point A on the curve. The problem is that the calculation of a gradient requires two points (an interval) but a tangent by definition is a line that grazes the curve at a single point. To calculate the gradient of the tangent exactly we must assign $\Delta x = 0$, but then we would no longer have an interval to calculate a gradient! Traditionally $\Delta x = 0$ is placed in (1) and the gradient of the tangent at A for the curve $y = x^2$ is given by (2) as

$$\frac{dy}{dx} = 2x \qquad (2)$$

where, the meaning of dy and dx now become unclear for they are no longer finite intervals. This difficulty is usually disguised by the meaningless statement "in the limit as Δx goes to zero". Either it goes to zero or it does not, and it does not because we are incapable of allowing it to do so. The correct conclusion should be that

$$\frac{dy}{dx} = 2x + \textit{very small number} \qquad (3)$$

which means that dx and dy are very small non-zero changes in x and y. This very small number can be made as small as one desires.

These considerations bring the whole issue at hand to a sharp focus. An interval can be subdivided into only a finite number and not an infinity of sub-intervals. Consider a straight line. If one places two identical sub-intervals P followed by Q in contact on it then either the higher limit of P must coincide with the lower limit of Q, or vice versa, thereby doubling the extent of the interval. This is what Aristotle has in mind when he states that:

> I say that something is continuous ... when the limits by which two objects [sub-intervals] are in contact have become identical and, as the word implies, enable one object to continue into the other.[3]

He recognizes that this is not possible for points:

> ... a line, which is continuous, cannot consist of points which are indivisible, first because in the case of points there are no limits to form a unity [sub-interval] (since nothing indivisible [like a point] has a limit which is distinct from any other part of it) and second there are no limits to be together [in contact] (since anything which lacks parts lacks limits too, because a limit is distinct from that of which it is a limit).[4]

In this light, Aristotle's proposed resolution of Zeno's paradox is puzzling.

> That is why Zeno's argument makes a false assumption, [in saying] that it is impossible to ... make contact with infinitely many things one by one in a finite time. For both length and time, and in general everything continuous, are said to be infinite in two ways – either infinitely divisible or infinitely long. If, then, they are infinite in quantity, then it is impossible to touch them all in a finite time; but if they are [only] infinitely divisible, it is possible to touch them – and in fact this is the way in which time is infinite[5]

However, what does Aristotle's "infinitely divisible" mean? Does he mean that a finite interval can be broken into an infinity of parts? If so, then Aristotle is admitting the possibility of a time interval possessing "infinitely many ... nows [points]" which implies infinitely many sub-intervals which can each have only zero extent (otherwise an infinity of continuous sub-intervals each having a non-zero magnitude would produce an interval of infinite extent). In fact, in Book III of his *Physics*, Aristotle repeats the notion of "infinitely divisible": "no actual magnitude can be infinite, but it can still be infinitely divisible".[6] Bertrand Russell evidently interpreted Aristotle's "infinitely divisible" to mean an infinite number of intervals of zero extent (points), and thought that an infinity of points could construct a finite interval:

> if Achilles ever overtakes the tortoise, it must be after an infinite number of instants have elapsed since he started. This is in fact true; but the view that an infinite number of instants makes up an infinitely long time is not true, and therefore the conclusion that Achilles will never overtake the tortoise does not follow.[7]

However, Aristotle has already said that "a line, which is continuous, cannot consist of points". A sub-interval has distinct lower and higher limits, and if we contract this sub-interval to zero, the lower and higher limits coincide into a point consisting of a single limit. If one now places two points p and q in contact then their zero sub-interval content ensures that the higher limit of p must coincide with the lower limit of q but also the lower limit of p must coincide with the higher limit of q and therefore any two points in contact become superimposed. If one now attempts to construct an interval from points in contact then the lower and higher limits of the interval will collapse to a single point. So a finite interval cannot consist of points.

Aristotle attempts to get out of his contradiction with the following statement:

So the reply we have to make to the question whether it is possible to traverse infinitely many parts (whether these are parts of time or of distance) is that there is a sense in which it is possible and a sense in which it is not. If they exist actually [infinitely many parts], it is impossible [to traverse them], but if they exist potentially it is not.[8]

What does Aristotle mean by "exist potentially"? We can either represent an infinitely small part (point) to our intuition or we cannot. The difficulty is that we cannot. As stated earlier, a point has limits which coincide thereby giving it zero content. However if we were to attempt to visualise it this way, nothing would appear in our imagination! In order to imagine a point we must assign it two arbitrarily close but distinct limits thereby elevating it to the status of a sub-interval. This permits it to have a content that we can represent to ourselves. However, it cannot then be a point! This brings out the rule that whatever we represent to ourselves must be constructed of sub-intervals. To each of these a finite magnitude must be assigned because only finite magnitudes can be registered in our human consciousness.

So the real upshot of Zeno's argument is the realization that humans are only capable of representing to themselves a finitely divided interval. This now brings us to an important distinction between the specification of a process and the actualization of that process.

Consider the infinite series given by (4).

$$\frac{1}{2} + \frac{1}{4} + \frac{1}{8} + \frac{1}{16} + \cdots + \frac{1}{2^r} + \cdots = \sum_{r=1}^{\infty} \frac{1}{2^r} \qquad (4)$$

The process involved is the adding of these diminishing fractions into a sub-total. Unfortunately, the infinity on top of the summation sign indicates that this process cannot be completed. Mathematicians claim that the sum is equal to 1 but actually this is impossible to demonstrate. First they multiply (4) by one half.

$$\frac{1}{4} + \frac{1}{8} + \frac{1}{16} + \frac{1}{32} + \cdots + \frac{1}{2^{r+1}} + \cdots = \sum_{r=1}^{\infty} \frac{1}{2^{r+1}} \qquad (5)$$

Then (5) is subtracted from (4) giving

$$\frac{1}{2} = \frac{1}{2} \sum_{r=1}^{\infty} \frac{1}{2^r} \qquad (6)$$

Multiplying both sides by 2 finally gives the sum equal to one.

$$\sum_{r=1}^{\infty} \frac{1}{2^r} = 1 \qquad (7)$$

However, there is a problem with this subtraction. There is no doubt that the early fractions cancel out but what happens higher up the series? This is far from clear, so to answer this, instead of the upper limit being infinity let it be a positive integer n. Equation (4) then becomes the finite series:

$$\frac{1}{2} + \frac{1}{4} + \frac{1}{8} + \frac{1}{16} + \cdots + \frac{1}{2^n} = \sum_{r=1}^{n} \frac{1}{2^r} \qquad (8)$$

After multiplying by one half we get

$$\frac{1}{4} + \frac{1}{8} + \frac{1}{16} + \frac{1}{32} + \cdots + \frac{1}{2^{n+1}} = \sum_{r=1}^{n} \frac{1}{2^{r+1}} \qquad (9)$$

Then we subtract (9) from (8) giving

$$\frac{1}{2} - \frac{1}{2^{n+1}} = \frac{1}{2} \sum_{r=1}^{n} \frac{1}{2^r} \qquad (10)$$

Multiplying both sides by 2 finally produces

$$\sum_{r=1}^{n} \frac{1}{2^r} = 1 - \frac{1}{2^n} \qquad (11)$$

At this point, an infinity is usually inserted in place of n in (11), the $1/2^n$ term vanishes, the sum equals 1, and the vague notion of "the limit as n approaches infinity" is introduced, but what does this phrase actually mean? It cannot mean that n actually *is* infinity because infinity is not a positive integer. Furthermore, no number however large can *approach* infinity for is not an object located anywhere on the real number line that is available to be approached. The German mathematician Carl Frederich Gauss (1777–1855), protesting in a letter to his friend's use of infinity, seemed to appreciate the misuse of infinity although he mistakenly thought that it can be approached:

> I protest against the use of infinite magnitude as something completed, which is never permissible in mathematics. Infinity is merely a way of speaking, the true meaning being a limit which certain ratios approach indefinitely

close, while others are permitted to increase without restriction. Infinity should only be used as 'a manner of speaking' and not as a mathematical value.[9]

So the $1/2^n$ term can *never* vanish no matter how large we choose our positive integer n, although we can be sure that the sum will always be less than 1. Consequently, the execution of our subtraction of (9) from (8) can never be completed resulting in the impossibility of maintaining that the sum in (4) has the value of 1. All we can sensibly demonstrate is that it is 'one minus some arbitrarily small value'.

A similar argument can also be applied to the notion of an infinite set which was explored by the German mathematician Georg Cantor (1845–1918). Cantor defined the "cardinality" of a finite set as the number of elements it contained (for example, {1, 3, 5} has cardinality three) and claimed that two sets can be shown to have the same cardinality if a one-to-one correspondence can be established between them. For example, since each element of the above set can be paired off with each element of the set {2, 4, 6} then we can say that they have the same cardinality. Cantor attempted to generalise this concept to infinite sets and claimed that if a one-to-one correspondence can be shown to exist between their elements then they possess the same cardinality (size). Consider the following example of a one-to-one mapping between the set of natural numbers and the set of even numbers:

1	2	3	4	5	...	n	...
2	4	6	8	10	...	$2n$	...

To Cantor, and to Galileo who also considered this particular problem, this meant that the two sets had the same size. However, this means that the odd numbers can be removed from the set of natural numbers and the remaining set of even numbers still has the same size as the set of natural numbers!

The problem with this kind of thinking has already been pointed out. To demonstrate a truth to oneself, it is not enough to give the process by which it might be obtained. One must be able to execute the process to completion and visually represent the result to oneself otherwise, how can one know whether or not the asserted truth is possible? The difficulty with the above example is that one can never complete the process of pairing up the natural numbers with the even numbers so no statement about the *totality* of an infinite set can ever be made. The concept of an infinite set has no meaning because it is impossible to construct it!

So notions of 'infinity' and 'infinitely divisible' are nothing but mathematical metaphysics, neither of which are capable of being imagined and are therefore meaningless. This is the real upshot of Zeno's paradox, his reference to the meaningless process of infinitely dividing a finite interval (division of a finite number by infinity). It is a task that Zeno speaks of but is one that neither he, nor anyone else, can complete. So here I propose a condition for a mathematical process to be meaningful:

> A mathematical process is only meaningful if it can be completed in a finite time by a machine of arbitrarily large but finite processing power.

This brings out the human character of mathematics as a set of rules representing the way humans think. However, these rules or processes must be verifiable in intuition. As reported by Russell, the German mathematician Gottfried Leibniz (1646–1716), who developed the differential and integral calculus independently of Newton, saw the human characteristic of error in the notion of the infinitesimal:

> He appears to have held that, if metaphysical subtleties are left aside, the Calculus is only approximate, but is justified practically by the fact that the errors to which it gives rise are less than those of observation.[10]

The preceding conclusions are similar to those reached by the Dutch philosopher Luitzen Egbertus Jan Brouwer (1881–1966) in his theory of Intuitionism,[11] an investigation into the foundations of mathematics developed out of Arthur Schopenhauer's demand that all concepts be based on sense intuitions. Brouwer saw mathematical proofs as an essentially human activity carried out in intuited time, and he held that the successful demonstration that an object with certain properties exists depends on our capacity to construct that object in a finite time. It is not enough to specify how the terms of an infinite series are to be generated to form a sum. If the sum cannot be completed in a finite time then we are unable to demonstrate its proposed value. The distinction made here is between a language that communicates how to execute a process, and our actual ability to execute that process in order to demonstrate its result to ourselves. However, if we cannot demonstrate it to completion then the specification is meaningless.

References

1. Russell, Bertrand. *Principles of Mathematics*, (Allen & Unwin: 1937), p.331.

2. Aristotle, (transl. Waterfield, R.,) *Physics*, VI, Chapter 9, 239, b14, Oxford World Classics (Oxford University Press: 1999).

3. *Ibid.*, V, Chapter 3, 227, a6.

4. *Ibid.*, VI, Chapter 1, 231, a21.

5. *Ibid.*, VI, Chapter 2, 233, a21.

6. *Ibid.*, III, Chapter 6, 206, a14.

7. Russell, B. "Mathematics and the metaphysics", in Newman, J.R., *The World of Mathematics* (Dover: 2003).

8. Aristotle, (transl. Waterfield, R.,). *Physics*, VIII, Chapter 8, 263, b3, Oxford World Classics (Oxford University Press: 1999).

9. Gauss, Carl Frederich, letter to Heinrich Christian Schumacher dated 12 July 1831, Werke, Bd. 8, p.216

10. Ref 1, p.325.

11. Van Stigt, Walter P., *Brouwer's Intuitionism*, Studies in the History and Philosophy of Mathematics, Vol. 2 (North Holland: 1990).

Mathematical Puzzles II

38. The Striking Clock

When Nutter Norman head-butted the grandfather clock one afternoon, after a row with his wife, the timepiece protested. While the minute hand resumed normal service, the hour hand immediately reversed direction at the same rate as it had gone forwards. When, Norman's wife, Wind-up Wendy, wound up the clock later that day, although the hour and minute hands were correctly positioned to read 6:35, the time that was displayed differed from the correct time by five hours. At what time did the clock begin strike action? [Solution 16]

39. Square Feet with Corn

Weedy Willie was getting too old to work the land alone so he decided to divide his corn field between himself and his four sons in proportion to their five work rates. He knew that Rastus, Wig, Twig and Swig together could plant a field of corn in five hours whereas Wig, Twig, Swig and himself together could manage the same task in six hours. So Weedy divided his field into a two-digit square number of parts and kept just one part for himself. Wig, Twig and Swig were identical triplets and each son received a whole number of parts. How many parts did each son get? [Solution 31]

40. The Broken Ruler

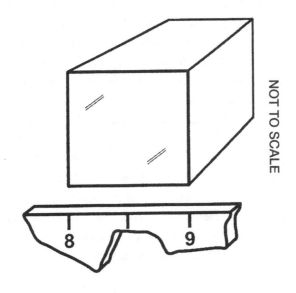

NOT TO SCALE

Mr Acrobat had to fill a cuboid-shaped packing box with a two-digit non-square number of identical cuboid-shaped chocolate boxes leaving no space in the larger box. The dimensions of a chocolate box (shown) are three consecutive whole numbers of inches, one of which he measured with a piece of broken rule that he had found (shown). More than one chocolate box spanned each of the three dimensions of the packing box which volume was a cube number of inches. What is the number of chocolate boxes that can be found in the smallest possible packing case volume (cubic inches)? [Solution 38]

41. Play on Words

At Gamblers Synonymous there was a game in progress. The words 'fear' (4 letters), 'alarm' (5 letters), 'afraid' (6 letters) and 'frightened' (10 letters) were written on four pieces of paper and surreptitiously sealed in four envelopes, one word to each envelope. Sid and Sally took it in turns to randomly choose an envelope, viewing their chosen word immediately after selection without showing the other person. The game ended when each had two words, and the winner was declared to be the one with the greater total of letters (e.g. 'alarm' and 'afraid' total 11).

In the game, Sally went first, but straight after her selection, she received inside information as to which envelope contained the word 'afraid'. Being the second largest word this was certainly to her advantage. She decided to select this envelope on her second turn should it still be available after Sid's first turn.

In playing this game many times, should her expected word total be less than Sid's, more than Sid's, or the same, and why? [Solution 34]

42. Fare Enough

Three Scotsmen, Swigger, Hughie and Ralph, were so drunk they decided that the best thing was for the three of them to share the cost of a cab. Now Hughie's home was eight miles from the pub, and when they eventually got him out of the taxi, he gave Swigger a £5 note. A mile up the road, the taxi reached Swigger's house. Swigger got out and gave Ralph a £20 note who returned a £5 note.

Ten miles from the pub, Ralph reached his house. He fell out of the taxi, gave the driver the £20 note and told him to keep the change. The fare was calculated at £1.80 per mile and each man should only have paid his share according to his distance travelled. Which of the four men made a profit and how much? [Solution 41]

43. Having a Ball

The Ball family ate so many sweets that each of the nine members had become a perfect sphere. Baby Ball was the youngest, with Toddler Ball twice Baby's diameter, Freda

105

Ball three times, Anne Ball four times, Kiki Ball five times, Eddie Ball six times, Rollie Ball seven times, Mama Ball eight times and Papa Ball nine times.

Three of the nine dived into the swimming pool. Just as they were climbing out, a fourth Ball dived in. Surprisingly, the water level of the pool when the fourth was fully submerged was exactly the same as when the other three had all been completely immersed. The fourth Ball was joined in the pool by two other Balls who had not previously got wet. The new water level when all were submerged was identical to the level after they had all climbed out and another Ball, who had not yet participated, had dived in and become fully submerged. Who stayed dry? [Solution 19]

44. Oliver's Didgeridoo

Oliver Hoot had just bought a didgeridoo. However, the sound seemed a bit muffled, and although the thing would didger it just wouldn't do. Back at the store he bought it from, the store owner discovered the problem: a cuckoo was lodging in the pipe. So Oliver picked it up and blew down one end to drive it out.

Now the three distances, of the cuckoo from each of the two ends and the pipe length, were all non-prime whole numbers (inches). Also the cuckoo would have taken the same time to fly to one end with the wind speed added to his natural speed as it would have taken to fly to the other end with the wind speed subtracted. It turned out that the cuckoo's natural speed divided by the wind speed was a square number less than 100, numerically equal to the pipe length. How long was Oliver's didgeridoo? [Solution 3]

45. Neddy's Workload

Neddy the workhorse has two carts to choose from, one red and one blue, each consisting of a square load-bearing area mounted on wheels. He usually pulls the maximum load he is capable of, so on one day he got the blue cart whose area is exactly covered with boxes of apples one layer high, and on another day he has the red cart exactly covered with boxes of oranges one layer high, the latter having an extra box of each fruit on top.

Although all fruit is stored in the same size cubic box, a box of oranges weighs a certain digit times the weight of a box of apples. Counted in boxes, the ratio of the single-digit length of the blue cart to that of the red cart is the same as the ratio of the sum of the weights of a box of each fruit to their difference in weight. How many boxes are on each cart? [Solution 23]

46. Armless Aliens

On planet Lim, the inhabitants are robots with various numbers of arms. The robots are grouped into tribes according to the numbers of arms they have so that in any tribe, no member has a different number of arms to the other members. Each tribe follows the same fertility ritual before bed each night, a ritual that also reveals the number of members in a tribe. In the ritual, each arm of a male links with an arm of a female, no male-female pair of robots linking more than once. The minimum number of robots required to link all arms once only is the number belonging to that tribe.

Now one night, the Creaky tribe cornered half of the Rusty tribe and despite being outnumbered two to one, unscrewed one arm from each, there being an equal number of male and female victims. However, the Creaky tribe were extremely poor at arithmetic and had failed to realise that they needed 24 times their actual haul of spare arms in order that the maximum number of links their members could then make, each with the same number of extra arms, was half the maximum number of links the Rusty tribe could then make. How many Creakies and Rusties were there in their respective tribes? [Solution 25]

47. The Slug and the Snail

A slug and a snail planned a race between two stones set a distance D cm apart. Said the slug to the snail "Since I'm slimier than you, I intend to start T seconds before you". "In that case", said the snail, who was rather slow, "I'm going to give you X cm start". It didn't matter. The slug, who could move at speed V cm/s, was only half as fast as the

108

snail. The result was a dead heat, t seconds after the slug set off. Now, D, t, T, V and X are all whole numbers from 1 to 10 inclusive, no two numbers being the same. How long did it take the snail to reach the finish? [Solution 52]

The Unexpected Hanging

In his article "The Paradox of the Unexpected Hanging", Martin Gardner[1] reports on a problem that was first published in 1948 by John O'Connor[2] who submitted it to the philosophical journal *Mind*. As Gardner relates it, a man is condemned to be hanged and, when sentenced on Saturday, the judge makes the following two statements: "The hanging will take place at noon on one of the seven days of next week" (Sunday to Saturday) and "You will not know which day it is until you are so informed on the morning of the day of the hanging". So we have the following:

(1) At noon on one of the next seven days you will be hanged.

(2) You cannot deduce which day the hanging will occur.

It is also given that "the judge was known to be a man who always kept his word". The prisoner's optimistic lawyer reasons as follows. The hanging cannot occur on Saturday, the last possible day referred to by the judge, because on Friday afternoon he would know that Saturday was the only day left and that the hanging was certain. This would violate (2) and so Saturday is eliminated. The argument continues that on Thursday afternoon, having eliminated Saturday by reasoning, Friday would be the only possible day, but this knowledge again violates (2) and so Friday is ruled out, and so on with all the other days. The lawyer concludes that the prisoner cannot be hanged, however on Wednesday the prisoner meets his fate. The problem is to find the flaw in the lawyer's reasoning.

To analyse the problem, we first need to simplify it to bring out the main points. We shall discover two problems, one concerning the contingent nature of predictions, and the other relating to the changing domain of applicability of the two conditions. These ideas will become clearer as we progress.

A Contingent Future

It was Monday morning, the weekend was over, and the students lazily drifted into their mathematics class. "At noon today," said the trustworthy Mr Tremble, "I intend to give you a test". Some children groaned, some exchange glances, and one even dropped his cell phone. "What's more," Tremble continued, "you shall be unable to deduce what day the test will occur". This time heads were scratched. He'd just told them that a test was to occur at noon so how could they not deduce there would be one at noon? Could their teacher carry out his intention accord-

111

ing to these conditions? Was Tremble going crazy? Let us number the statements as follows.

(3) At noon today there will be a test.

(4) You cannot deduce what day the test will occur.

What does it mean to say that 'before the test, (3) must be true'? Is it possible that a man as trustworthy as Mr Tremble, a mathematics teacher with a distinguished forty-year career, would not carry out his word? What if we specify that he *must* deliver his promise in (3)? Unfortunately, we now reach a dilemma. The point is that the students could never *know* in advance that his intention as specified in (3) will *necessarily* be carried out. For the eighteenth century philosopher Immanuel Kant (1724-1804), (3) is an *a posteriori* statement, one that depends for its truth on experiences yet to occur. Such pronouncements are characterized by their contingency, that is to say, contrary to mathematical statements that are *a priori* or necessarily true, something could happen to prevent it. Perhaps the fire bell will sound or Mr Tremble will be taken ill. One could conjure up all manner of calamities, but however unlikely they are, in an unpredictable world there is nothing to guarantee that they will not happen. No matter how sincere Mr Tremble is, the possibility cannot be eliminated that he will be obstructed in his intention and this reliance on a contingent future creates doubt. So the truth of (4) appears to rest on the uncertainty of the execution of (3).

There is something rather discomforting in this analysis. At first sight it appears to be a problem in pure logic, a province in which experience has no place, yet it somehow makes an appearance. So our first task is to eliminate the uncertainty in (3) resulting from possible external influence.

To do this, we turn to a related problem, one that removes all consideration of a contingent future from the analysis. The Paradox of the Unexpected Egg was devised by Michael Scriven[3] and here we consider a reduced version.

An assistant is asked to open a box with the following conditions.

(5) The box has an egg in it.

(6) At no point before opening the box you be able to deduce that it contains an egg.

Should the assistant expect to find an egg in the box? Contrary to the Unexpected Hanging problem, it is easier to argue that the existence of the event 'finding an egg' is unhindered by an external interaction. In the Unexpected Hanging, the production of the test lies in the future, whereas here the egg has already been secreted in the box. This makes (5) *verifiably* true and the participant reasons that this violates (6) which claims that no such confidence is possible. This violation itself creates uncertainty thereby validating (6) but that is because (5) and (6) are inconsistent rules for the given procedure of opening the box. There is something missing from this specification, something that when included should immediately underline our conclusion, and to discover what it is, we now take a sidestep into mathematics.

A Mathematical Problem

An assistant is invited to choose any whole number, which we shall call x, so that it satisfies the following two rules.

(7) When 1 is added to it, the answer is less than 5.
(8) When 1 is subtracted, the answer is greater than –1.

He can then use that number to select an identically numbered box. His choice might be $x=5$ so that (7) results in the answer 6 and (8) gives 4. Condition (8) is satisfied but (7) is violated so the opening of a box cannot be effected. The problem is, the rules of the game can only be satisfied when the whole number x is greater than 0 and less than 4, in other words, for 1, 2 and 3. Mathematically, this is called the 'domain', which consists of the region of applicability of the rules, and it can be written $0 < x < 4$, . Once this domain is stated correctly, thereby restricting the assistant's choices, the game can always proceed.

An Extended Egg Problem

We now return to the egg problem, but extend it to two boxes, borrowing the feature from the mathematical problem that the domain of the rules must be specified.

An egg is concealed in only one of two boxes labelled 1 and 2. The assistant is asked to open the boxes sequentially (that is, 1 then 2) with the following conditions.

(9) One and only one of the boxes contains an egg.
(10) At no point in the procedure is it possible to deduce which box contains the egg.

Domain: Boxes 1 and 2

The assistant notes that domain states that the rules are valid for box 1. Before opening box 1, he knows that the egg is confined to only one of two boxes and that he has insufficient information to deduce which one. Here, both rules (9) and (10) are consistent. He also notes that the domain states that both rules apply to box 2 so he starts to reason ahead. He concludes that if there is no egg in box 1 then it must be in box 2. However, this would violate (10) so he concludes that it cannot be in box 2, and to satisfy (9) it must be in box 1. This would again violate (10) and so there is no egg in any box. This now violates (9) and the assistant is by now thoroughly confused. He decides to proceed anyway, he opens box 1, and finds no egg. He opens box 2 and there it is! Where was the fault in his reasoning?

Let us return to the point in the game after the opening of box 1, which has no egg, and before the inspection of box 2. The assistant deduces from (9) that the egg must be in box 2 but this now violates (10). The error lies in his conclusion that this violation constitutes a deduction that the egg is not in box 2 but what it actually means is that the domain incorrectly specifies that the rules are consistent for box 2. They are not. When he is faced with a choice between two boxes there is no problem with (10), but it breaks down when only box 2 remains. It is a game that cannot be executed because the rules are not consistent in all parts of the domain. Fortified with the results of our analysis, we return to Mr Tremble's promise of a test but extended over a two-day domain.

The Two-Day Domain

On Tuesday, Mr Tremble takes a different class and informs them that it is his intention to give them a test that week at noon on "either Thursday or Friday". This time there are no groans, the group evidently being more mature. "However," Tremble added, "you shall not be able to deduce in advance which day you shall receive it". It was too good to be true! Let us take ourselves to Friday morning, with Thursday having passed without a test being given. We label the statements as below.

> **(11) At noon on either Thursday or Friday there will be a test.**
>
> **(12) You cannot deduce what day the test will occur.**
>
> **Domain: Thursday and Friday**

If we arrived at Friday morning without a test being given, the domain of validity of the rules compels us to create the idealization that the test must occur at Friday noon. This violates (12) and it is a violation that shows that the rules are inconsistent for Friday and that the domain is incorrectly given. This situation exactly parallels the Paradox of the Unexpected Egg.

The Unexpected Hanging

It is now clear why the lawyer's reasoning is faulty. First, the situation is obscured by the contingent nature of rule (1) which does not prevent external influences from invalidating the intent of rule (1) and validating rule (2). However, once the domain is specified (Sunday to Saturday) then the game is idealized and all such influences can be removed from consideration. It is now certain that a test must occur somewhere in this domain. Once this difficulty has been removed, a second problem arises in that the 'game' cannot be executed over all parts of the domain, in particular, the last day Saturday. In this circumstance, the conclusion should not be that the test cannot occur on Saturday but that the rules are inconsistent on that day. Here, no deduction is possible and this amounts to producing no valid information at all. So my conclusion about the Unexpected Hanging is that the problem has an ill-defined domain.

References

1. Gardner, M. "The paradox of the unexpected hanging", *The Unexpected Hanging and Other Mathematical Diversions*, (1969).

2. O'Connor, D.J. "Pragmatic paradoxes", *Mind*, **57**: (1948), 358–9.

3. Scriven, M. "Paradoxical announcements", *Mind*, **60**: (1951), 403–7.

The Wave-Particle Puzzle

The French philosopher Pierre Gassendi (1592–1655) was the first to apply the atomist theory to light, a view that was subsequently developed by Isaac Newton (1642–1727). However, in 1801, the English physicist Thomas Young (1773–1829) passed light through two closely spaced pinholes and noticed that when the two resulting beams fell together on a screen they created an alternating pattern of light and dark bands. This phenomenon of interference strongly suggests that light is wave-like. When the distances that the two beams travel from the pinholes are such that their peaks coincide on the screen, then they create a bright fringe, and when the two distances are such that the peak of one meets a trough of the other, then they cancel out as a dark region. In the early 1860s, James Clerk Maxwell's (1831–1879) success with his wave theory of electromagnetism further suggested that the energy of a light photon is continuously distributed over an increasing volume. Although this successfully accounted for optical phenomena, in 1905 Albert Einstein (1879–1955) asserted that this concept could not explain the phenomena of blackbody heat radiation, photoluminescence, and the ultra-violet production of cathode rays. Instead he suggested that a light photon must be:

> localized in space ... [must] move without being divided [and is] ... absorbed or emitted only as a whole.

In *Quantum Physics: A First Encounter*, Valerio Scarani, a Senior Researcher at the University of Geneva, Switzerland, describes a series of experiments, increasing in complexity, that clearly illustrate the conceptual difficulty inherent in assigning either a wave or a particle nature to matter and light. Each experiment presents a source of emissions together with a number of strategically placed beam splitters and detectors. While the prevailing view is that the nature of the beam is particle-like, we shall refrain from such a commitment and shall instead refer only to emissions.

The first experiment (Scarani 4–6, see Figure 1) sends a low intensity light beam towards a semi-transparent mirror which acts as a beam splitter.

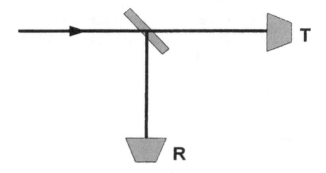

Figure 1. One beam splitter and two detectors.

Each of the two paths out of the beam splitter is directed towards a detector. After a large number of registrations at the detectors the following observations are noted.

(1) The two detectors do not register simultaneously.
(2) Half the registrations are at T and half at R.

The interpretation given is that "a particle arriving at the beam splitter is not divided but it is either transmitted (T) or reflected (R)".

119

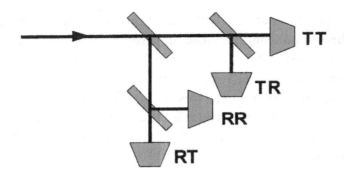

Figure 2. Three beam splitters and four detectors.

The second experiment (Scarani 6–7, Figure 2) passes each of the paths out of the beam splitter through another beam splitter to produce four paths, each of which reaches a detector. According to Scarani, the "particle" may either be transmitted twice (TT), transmitted at the first splitter and reflected at the second (TR), reflected at the both splitters (RR), or reflected at the first and transmitted at the second (RT). This time, 25% of the registrations occur at each detector.

The third experiment (Scarani 7–8) introduces the notion of interference through the use of a 'balanced Mach–Zehnder interferometer'. Each of the two paths exiting the first beam splitter in Figure 1 is reflected at a mirror towards a second common beam splitter with two exit paths, each met by a detector (Figure 3).

The detectors are labelled according to the outcomes at the two beam splitters resulting in four possible paths with equal length.

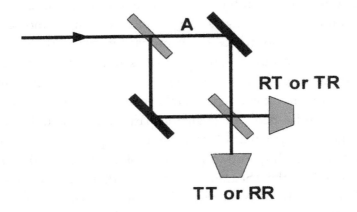

Figure 3. Interference with two splitters and two detectors.

On the basis of the first two experiments one might expect that 50% of the registrations occur at each detector, however, it is observed that all of the registrations occur at the "RT or TR" detector. According to the wave theory, if we assume that each beam splitter and reflector advances the phase of the wave by a quarter cycle after reflection, then the following results. There will be two in-phase waves at the "RT or TR" detector, each with a half-cycle advance, so that they constructively interfere. The two waves at the "TT or RR" detector have advanced by a quarter and three quarters of a cycle, so they arrive a half cycle out of phase and destructively interfere. This accounts for the result. The wave theory also accounts for the observation that if the emission component at A in Figure 3 is arrested, then 50% of the registrations occur at the two detectors, that is, the interference disappears. The difficulty with the wave theory lies in the circumstance that in the first experiment, on no occasion do the detectors register simultaneously, discouraging the interpretation that more than one path is simultaneously traversed by a single emission, and presenting the appearance of an indivisible and particle-like structure. Scarani enquires "How is it that the particles that travel by the path that we have not modified can know about the modification [at A]?" concluding that "each particle is 'informed' about all the paths that it could take, without being actually 'divided' into two parts" (Scarani 9) and that "each quantum particle explores all the indistinguishable paths" (Scarani 25). It seems that the acceptance of this mystical 'explanation' has both diverted attention away from the conceptual crisis and assisted in obstructing further progress.

I should like to suggest a way out of this dilemma and it is connected with the assumption that underpins the above interpretation of these observations. This assumption, the validity of which seems not to have been challenged hitherto, is that each arrival at a detector is *necessarily* accompanied by a registration. In fact, we do not yet know whether or not an emission component can pass through a detector unregistered for it would require a registration at one of a series of consecutive detectors placed further along the same path to demonstrate it. Whether or not the number of these semi-permeable detectors would be too large and the observation time too long to prohibit the assembly of such an affirmative experiment is an open question. Nevertheless, our alternative hypothesis would indeed permit an emission to divide at the beam splitter in Figure 1 without the observation of simultaneous registrations at the detectors. It would imply that the conditions for an emission component to affect a target atom are not always met. As an emission passes through the system, the probability of one detector being affected could be so small as to render the observation of two simultaneous registrations exceedingly unlikely within the running time of the experiments so far conducted. Without a registration necessarily accompanying an arrival at a detector, an accurate count of the number of emissions passing through the apparatus per unit time becomes an unreachable upper limit, for in addition to being either absorbed for registration or transmitted without registration at a detector, an emission might also be scattered and thereby lost to the possibility of detection.

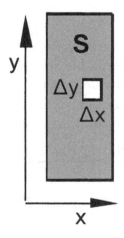

Figure 4. Element of area ΔxΔy on surface S.

Consider the surface of a detector with area S upon which a homogeneous monochromatic light wave impinges (Figure 4). We let the source run for a time period Δt, sufficiently large to allow the registration of a significant number of absorptions n. Contrary to Max Born, we now assume that the wave does not represent the probability that a point-like photon can be found at a certain location, in other words, it is not merely a calculating device that is devoid of substance, but that it is has material content. If all atoms in the surface were available to repeatedly absorb a part of the wave then the total number of absorptions in area S in unit time would be N, a number beyond practical determination. However, I posit that in time Δt not all atoms are disposed to absorption, and that only n registrations per unit time can actually occur, where $n \ll N$. We now write the probability that an absorption A occurs on surface area S as

$$P(A) = \frac{n\Delta t}{N\Delta t} = \frac{n}{N} \qquad (1)$$

Let the n observed absorptions take on a non-homogeneous distribution across the surface S so that the number of registrations per unit area in unit time is given by the function $f(x,y)$. The probability of an observation O in $\Delta x \Delta y$, given that an absorption A is registered on S, is then

$$P(O|A) = \frac{f(x, y)\Delta x \Delta y \Delta t}{n\Delta t} = p(x, y)\Delta x \Delta y \qquad (2)$$

where $p(x,y)$ is the probability per unit area. Since

$$P(O \cap A) = P(O|A) \bullet P(A) \qquad (3)$$

Then

$$P(O \cap A) = \frac{f(x, y)\Delta x \Delta y}{n} \bullet \frac{n}{N} = \frac{f(x, y)\Delta x \Delta y}{N} \qquad (4)$$

Here, $P(O \cap A)$ denotes the probability of both an observation and an absorption, which might be exceedingly small if N is very large. Hitherto, n/N has been taken to be unity so that has appeared to be much larger than the present theory suggests. On the contrary, if is exceedingly small, the probability of two simultaneous absorptions on S could be negligible, even under the circumstances where the time period Δt is large and the function $p(x,y)$ is a maximum. These considerations allow discrete absorptions to be entirely consistent with a wave theory of radiation.

Of course, this leaves the question as to the nature of the "conditions for an emission component to affect a target atom". If these are not satisfied then the wave-front will pass by a surface atom of the detector. I would like to suggest that a target atom requires a favourable orientation in relation to the incident wave for an electron to be excited, and that it is this randomness in orientation of the target atoms in the detector surface that accounts for the randomness of absorption sites. With this view, Einstein's concept of the photon as 'localised' would have to be given up in favour of a theory of fluid motion in which a minimum unit of action for an electron would result from the gradual accumulation of fluid in an atomic system. Once this action is fully resident then an atomic electron as a quantity of fluid becomes available to participate in emission and absorption processes. Such an absorption and emission could create a directed chain of these processes, each atom in the chain being saturated by the preceding emission, and itself emitting to contribute to a single observable track. These two considerations, orientation and the accumulation of action, might well be the key to understanding the wave particle dilemma, and the resulting fluid theory of matter would provide grounds for an explanation as to how a mass can affect another mass at a distance.

Creative Thinking Puzzles III

Each puzzle has two hints which can be found at the back of the book.

48. Moving Clocks

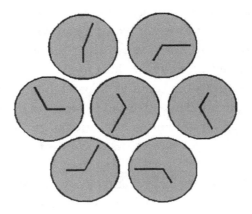

Shown are seven clocks showing different times. Can you rearrange them among the seven positions, without changing their orientation, so that they all show the same time? [Solution 24]

49. Miserable Marriage

A woman's husband was also stepfather to her children. Unfortunately, it was a loveless marriage and after describing him in her diary as shown she decided to separate. Can you remove one of the letters to assist her in leaving her husband? [Solution 58]

50. Out of This World

If you found an extra-terrestrial being like the one shown what would he be in? [Solution 46]

51. Doubtful Date

A man married his girlfriend on the date shown. If her minimum age had to be 18 how old was she? [Solution 35]

52. Door to Door

In Correspondence Close, exactly two of the three house numbers shown got to equal first place in a letter writing competition. What are the two numbers? [Solution 21]

53. No Earthly Connection

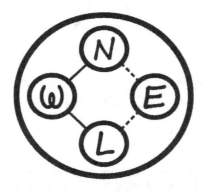

After his recent visit to Earth, Alpha the alien incorrectly drew the above. Nothing needs to be added to or removed from the drawing to correct it. Can you change the content of one of the small circles to rectify the picture? [Solution 55]

54. Water Puzzle

Mervin the magician has just turned water into wine. However, none of his audience at the Teetotal Club is particularly impressed. By relocating the cocktail stick and placing it horizontally, can you reverse the effect and change the wine into water? Can you also find a second solution by rearranging the stick vertically? [Solution 50]

55. Doing a Turn

Shown are three smokers performing a song on stage. The whole number is performed as a three-part harmony.

Q: What is it? [Solution 47]

56. The Tin Door

While exploring a cave, two adventurers unexpectedly found a locked tin door set into a wall. On the adjacent wall was a set of curious fractions. Can you decipher the message and discover what is in the secret room? [Solution 33]

57. Fish Feast

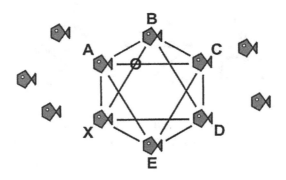

A shoal of fish were enjoying a swim when the fish at X decided to eat the five fish A–E. This it did by moving only along the straight lines, visiting each position once only and finally returning to X to swim along with the shoal. Now C was eaten some time before D who was not the last eaten, B was devoured some time before A and the route that was taken did not cross over itself at O. Can you draw the route that fish X took? [Solution 53]

58. Dig It

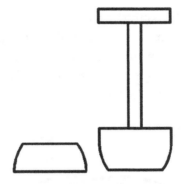

Shown is a stone, to the right of which is a spade stuck in the ground. Can you pick up the stone and add one straight line to the picture to show that the spade has sunk deeper into the ground? [Solution 60]

59. The Concealed Car

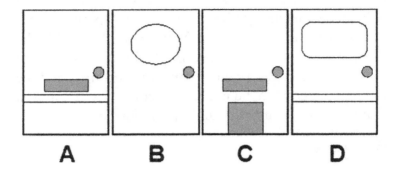

Behind one of the four glass doors is concealed a car. Door A has a letterbox and a push-bar, B shows an oval window, C exhibits a letterbox and a cat-flap, and D has a rectangular window and a push-bar. Which door is to be opened to reveal the car? [Solution 48]

60. Pet Theory

Can you add the four straight lines at the bottom of the picture to the shapes above them to complete the view of two identical pets? [Solution 37]

Titan's Triangle

There is a problem that often turns up in IQ tests that involves finding the number of shapes in a larger shape. An interesting example is shown below where the number of triangles of any size is to be counted in a triangle with side length four ($n = 4$).

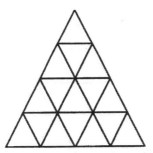

Titan's Triangle with n=4

A generalisation is provided here for a triangle of any size n, in which formulae are given for the total number of triangles, parallelograms, and trapezoids of any size. Initial results are also presented for the number of ways closed paths of length 4, 5, and 6 might be traversed, by focusing on specific shapes. The reader is invited to take up the project and make new discoveries of his own.

Number of shapes

If one wishes to find a formula for the sum of triangles S_n of any size in a triangle of side n, then care must be taken in applying a mathematical technique such as the calculus of finite differences because there are two cases, one for n even and the other n odd.

133

$$S_n = \frac{1}{8}[n(2n+1)(n+2) - \delta_n]; \quad \delta_n = \begin{cases} 0 \; for \; n \; even \\ 1 \; for \; n \; odd \end{cases}$$

A calculation at $n = 4$ verifies that the answer to the problem given above is 27 (did you include the inverted triangle of size 2?). An arbitrary number of layers can be added to the triangle (the following figure shows an $n = 7$ triangle).

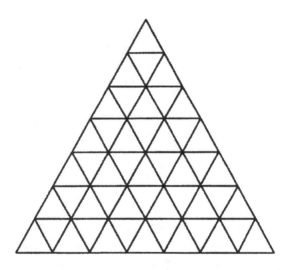

Titan's Triangle with n = 7.

As the triangle grows some interesting problems arise. For example, we need not confine ourselves to counting triangles but could extend our investigation to quadrilaterals. Two types occur in our triangle: parallelograms and trapezoids. We shall designate a parallelogram by two numbers which give the side lengths (a, b) where $a \geq b$. The trapezoid is also to be described by two numbers (c, d) where, in relation to the main triangle, c is the length of the innermost parallel, d the length of the outermost, and either could be the bigger. This description also applies when the trapezoid is rotated 60 and 120 degrees.

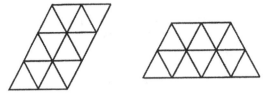

Parallelogram (3,2) and trapezoid (2,4).

In Tables 1 and 2 we show the number of parallelograms of dimensions (*a, b*) in a triangle of side *n*.

	1	2	3	4	5
1	45	60	36	18	6
2		18	18	6	
3			3		

Table 1. n = 6, column a, row b

	1	2	3	4	5	6
1	63	90	60	36	18	6
2		30	36	18	6	
3			9	6		

Table 2. n = 7, column a, row b.

135

We note that a is the column number and b the row number. The formula for predicting the elements $(a, b)_n$ in these tables, for example $(3, 2)_7 = 36$, is as follows:

$$(a, b)_n = \frac{3[n + 1 - (a + b)][n + 2 - (a + b)]}{\delta_{ab}}$$

$$a \geq b; \qquad \delta_{ab} = \begin{cases} 1, & a \neq b \\ 2, & a = b \end{cases}$$

$$a = b; \quad \begin{cases} b = 1, 2, 3, \ldots, \dfrac{n}{2}, & n \text{ even} \\ b = 1, \ldots, \dfrac{n - 1}{2}, & n \text{ odd} \end{cases}$$

$$S_n = \frac{1}{8} n(n^2 - 1)(n + 2)$$

In contrast, using these tables, the sum S_n of all the parallelograms in a triangle of size n is remarkably simple.

	1	2	3	4	5
1		30	9		
2	45		18	3	
3	30	30		9	
4	18	18	18		3
5	9	9	9	9	
6	3	3	3	3	3

Table 3. No. trapezoids at n = 6, column c, row d

In Tables 3 and 4 we give the number of trapezoids of dimensions (c, d) in a triangle of size n. The elements T_{cd} of the table adhere to the following rules.

	1	2	3	4	5	6
1		45	18	3		
2	63		30	9		
3	45	45		18	3	
4	30	30	30		9	
5	18	18	18	18		3
6	9	9	9	9	9	
7	3	3	3	3	3	3

Table 4. No. trapezoids at n = 7, column c, row d

There are two cases, first for $c < d$:

$$T_{cd} = \frac{3(n + 1 - d)(n + 2 - d)}{2}$$

$$c = 1, \dots, n - 1; \quad d = c + 1, \dots, n; \quad c < d$$

So for $n = 6$, $c = 2$, $d = 3$ we have

$$T_{23} = \frac{3(4)(5)}{2} = 30$$

The other case is for $c > d$:

$$T_{cd} = \frac{3(n + 1 - 2c + d)(n + 2 - 2c + d)}{2}$$

137

$$\Delta = c - d = 1, 2, \ldots, q; \qquad q = \begin{cases} \dfrac{n-1}{2} & \text{for } n \text{ odd} \\ \dfrac{n}{2} - 1 & \text{for } n \text{ even} \end{cases}$$

$$c = \Delta + 1, \ldots, n - \Delta; \qquad d = c - \Delta$$

For example, for $n = 6$, $c = 3$, $d = 2$, we have

$$T_{32} = \frac{3(3)(4)}{2} = 18$$

However, of greater interest is the trapezoid formula that predicts the sum S_n of the numbers in the table for any n:

$$S_n = \frac{n(n^2 - 1)(n + 2)}{8}(1 + P)$$

$$P = \begin{cases} \dfrac{n(n-2)}{2(n^2 - 1)} & \text{for } n \text{ even} \\ \dfrac{n^2 - 3}{2n(n + 2)} & \text{for } n \text{ odd} \end{cases}$$

If we disregard the trapezoids above the leading diagonal of the table where $c > d$, in other words we miss out all trapezoids where, of its two parallels, the one on the inside of the triangle is bigger than the one on the outside, then we reach exactly the same formula as that obtained for the total number of parallelograms (P=0). Combining our two results for the parallelograms and both types of trapezoid, we have the following:

$$S_n = \frac{n(n^2 - 1)(n + 2)}{8}(2 + P)$$

where P is defined as above. This is the formula for the total number of quadrilaterals in a triangle of size n.

Number of ways of traversing a closed path of length m

Our problem is now to find relations that predict the total number of ways N_{mn} that a closed path of length m can be traversed within a Titan's Triangle of size n. Some of the results of the previous analysis will undoubtedly assist in this task.

The number of closed contours of length $m = 3$ in a triangle of dimension n is the number of triangles of side length one which is given by

$$N_{3n} = n^2$$

For closed contours of length $m = 4$, we already have the number of parallelograms where $a = b = 1$:

$$N_{4n} = (1,1)_n = \frac{3n(n - 1)}{2}$$

A contour length $m = 5$ only occurs for the trapezoids T_{12} and T_{21} for which the formulae have already been found:

$$T_{12} = \frac{3n(n - 1)}{2}$$

and

$$T_{21} = \frac{3(n - 1)(n - 2)}{2}$$

which when added gives

$$N_{5n} = 3(n-1)^2$$

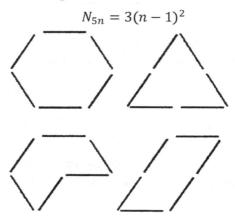

Allowed contours for m = 6

However, when we get to $m = 6$, as well as vertices that point outwards we also get ones that point inwards. If we constrain the possible shapes so that no point may be passed through more than once, for this would produce more than one distinct region, then the allowed contours are as shown.

The number of hexagons of contour 6 in a triangle of side n is as follows:

$$H_{6n} = \frac{(n-1)(n-2)}{2}$$

The third shape is a hexagon with a cut-out and has six rotations so that altogether we need $7H_{6n}$. The second shape is a triangle of side 2 which total is given by $(n-1)(n-2) + 1$. The final shape is a parallelogram which we have previously calculated to have a total:

$$(2,1)_n = 3(n-1)(n-2)$$

140

Adding all these gives

$$N_{6n} = \frac{13}{2}(n-1)(n-2) + 1$$

The problem is really an analysis of travel through a particular type of network in a closed path consisting of a given number of straight lines. The formulae give the number of ways this can be done. Of course, further results could be developed and the interested reader might be inspired to extend the contour formulae to $m > 6$. For example, does $m = 7$ follow the above pattern and become

$$N_{7n} = A(n-2)^2 + B$$

where A and B are constants?

CREATIVE THINKING
HINTS

Creative Thinking Puzzles I: First Hint

11. Mad House
This is not actually a house.

12. Right Angle
The title is really an instruction.

13. Amazing
Try a different type of path.

14. The Dead Dog
The legs do not remain legs.

15. Sum Line
Is it really a 9?

16. Which Way?
View all five circles together.

17. Nothing for It
Rearrange includes rotate.

18. Rough Graph
This is not a graph.

19. The Lighthouse
The three lines of the sail could be three wires.

20. The Pig and the Bird
What parts of the picture might belong to a bird?

Creative Thinking Puzzles I: Second Hint

11. Mad House
Try a different angle.

12. Right Angle
You need to use your head.

13. Amazing
Think negatively!

14. The Dead Dog
The mouth is not the mouth and the eye is not the eye!

15. Sum Line
You could try standing on your head!

16. Which Way?
Let us think in black or white.

17. Nothing for It
The letter N is not a letter N.

18. Rough Graph
There's a sound in the title.

19. The Lighthouse
How might we demonstrate an identical load?

20. The Pig and the Bird
It's staring straight at you!

28. Inspector Lewis
Why are there gaps in the picture?

29. The Builder's Problem
The sand need not sit horizontally in the box.

30. The Four Dice
Why are there thick and thin lines?

31. Politically Correct
Some biology might help.

32. Inklined
How can tipping the bottle reveal the time?

33. Bubble Math
Try adding straight lines.

34. Arithmystic
The problem is more English than mathematics.

35. No Escape
The solution has no letter box in it.

36. Secret City
The city is abbreviated.

37. Winning Line
There is a way that some letters might be seen as different ones.

Creative Thinking Puzzles II: Second Hint

28. Inspector Lewis
There are five distinct parts.

29. The Builder's Problem
A slight turn of the head might help.

30. The Four Dice
Each die is something else.

31. Politically Correct
Think male and female.

32. Inklined
A digital clock might help.

33. Bubble Math
A clue in is in the name.

34. Arithmystic
Does it need spelling out for you?!

35. No Escape
Where can the knob be placed to show the door open?

36. Secret City
The secret is in MY CITY.

37. Winning Line
The added underline completes one of the letters.

Creative Thinking Puzzles III: First Hint

48. Moving Clocks
Only one of the clocks shows the eventual time.

49. Miserable Marriage
The letter A could be the side view of some object.

50. Out of this World
Short name for extra-terrestrial?

51. Doubtful Date
Try an alternative way of writing the number 17.

52. Door to Door
'First place' refers to the numbers of the houses.

53. No Earthly Connection
Think electricity.

54. Water Puzzle
Think chemistry.

55. Doing a Turn
Could you do a turn?

56. The Tin Door
There is a reason that the door is made of tin.

57. Fish Feast
What do the possible routes look like?

58. Dig It
The stone must first be removed.

59. Concealed Car
Parts of the car are already visible.

60. Pet Theory
One of the two pets is partly behind the other.

Creative Thinking Puzzles III: Second Hint

48. Moving Clocks
How can a clock be constructed?

49. Miserable Marriage
One letter appears three times.

50. Out of this World
The alien is constructed from four parts.

51. Doubtful Date
Think partitions.

52. Door to Door
Try writing out the house numbers.

53. No Earthly Connection
Altering one of the letters produces an electrical component.

54. Water Puzzle
What is the picture constructed from?

55. Doing a Turn
The conundrum is asking about Q.

56. The Tin Door
The message can be solved by not looking directly at it.

57. Fish Feast
After the meal, the fish is swimming with the shoal.

58. Dig It
How might the stone be used to sink the spade in the ground?

59. The Concealed Car
How can the parts of the car be combined?

60. Pet Theory
You'll never need to give your pets a drink.

SOLUTIONS

1. Digital Dilemma. The unique solution is 21475 which when, each digit is replaced by the letter in that alphabetic position, spells BADGE.

2. Mad House. If one rotates the picture a quarter turn anti-clockwise we see a man smoking a cigarette. So (c) relaxed is correct.

3. Oliver's Didgeridoo. The didgeridoo was 49 inches long. Let the short and long distances of the cuckoo from each end be x and y, respectively. Let the cuckoo speed be v and the wind speed be u. Then the equal times relation is

$$x/(v - u) = y/(v + u)$$

which by rearrangement gives

$$y - x = (u/v)(x + y) = 1$$

taking into account that $v/u = x + y$. The only non-prime positive integer values of x and y less than 100 which differ by 1 and sum to a square are $x=24$ and $y=25$ so that $x + y = 49$.

4. Which Way? Add a white arrow pointing downwards to make a letter U so that the sign reads SOUTH. (The hint 'Let us' means 'letters'!)

5. Word in the Stone. The English word was JADE (the French word was DEJA). Let the four numbers be W, X, Y, Z and their total be T. Then we have $(W + X + Y)/Z = (T - Z)/Z$. Since the left hand side of the equation is a whole number then $(T/Z) - 1$ and hence T/Z are also whole numbers. By symmetry, W, X, Y, Z are factors of T. The possible factors of 20 are 1, 2, 4, 5, 10, 20. The four of these that sum to 20 are 1, 4, 5, 10. Replacing these by letters gives A, D, E, J.

6. Safe Cracker. 8425

7. Santa Flaws

	Dwarf	Reindeer	Gift	House
1	Doc	Prancer	MP3 Player	Binsleepin
2	Bashful	Dancer	Radio	Litesout
3	Sleepy	Donner	Doll	Nomunie
4	Dopey	Blitzen	Guitar	Takerhike
5	Sneezy	Cupid	TV	Dungroovin
6	Happy	Comet	Computer	Whywurry
7	Grumpy	Vixen	Book	Ronguns

8. Sum Line

9 = 7 + 8 − 6

Remove the vertical line from the 4 and view upside down.

9. Maximum Security. The volume of the cell cannot exceed 24. The temptation is to write down the volume of the cell by multiplying the three sides as

$$V = -x^3 + 6x^2 - 11x + 6$$

differentiate V with respect to x, then set it equal to zero to find the location of the maximum volume. However, at least one of the cell walls will then have a negative value for these two x

solutions. Furthermore, the volume at this x location will not be the greatest volume possible ($V \sim 0.38$). Since the window must exist and no wall length can be negative then $-1 < x < 1$. Substituting this in the volume equation gives $0 < V < 24$. This brings out the point that the maximum value of a function at a turning point is not necessarily the highest value of the function.

10. The Five Chimneys. The smoke directions are A(+), B(−), C(+), D(−), E(+). There are only two possible continuous circles of relations that can be formed that avoid contradictions. One of them is given as the solution and the other is A(−), B(+), C(−), D(+), E(−). The latter does not constitute a positive total of smoke and so is eliminated. In case anyone is unclear whether or not there really is a Beijing district of Foo Mee Gate, one can substitute letters for digits (indicating alphabetic position) in the house number 692 to arrive at the word FIB!

11. Alien Mutations A body circle, B remove front appendages, C remove rear appendages, D remove antennae, E to quadruped, F head circle, G body circle, H add antennae, I remove rear appendages, J add front appendages, K remove front appendages, L body square, M add rear appendages, N to biped, O add antennae, P add front appendages.

12. The Lighthouse

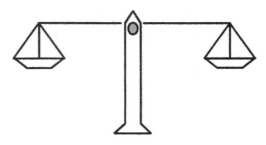

The solution shows weighing scales.

13. Winning Line

If the diagram is viewed upside down, adding an underline turns the V into an A.

14. Core Conundrum

1	4	5	8
3	9	1	5
9	2	4	3
5	3	8	2

15. Politically Correct

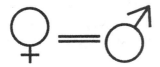

16. The Striking Clock. The clock began strike action at 3:05pm. A valid time – where the hour hand has the correct position for the minute shown – only occurs every half hour after the reversal of the hour hand starts, so at any valid time the clock can only appear a whole number n hours fast or slow. In a 12-hour period, if the clock appears n hours fast, then the initial time was either $6 - n/2$ or $12 - n/2$ hours ago, and either $n/2$ or $6 + n/2$ hours ago for n hours slow. This amount is added to the reversed time to get the initial time. For 6:35 and $n=5$, we have the following possible initial times: 10:05 (3.5 hours ago), 4:05 (9.5), 9:05 (2.5) and 3:05 (8.5). Neither the first nor the third are in the afternoon and if the second is, then the clock is wound the next day (invalid) at 1:35am. The final case gives a wind-up time of 11:35pm.

17. Rough Graph

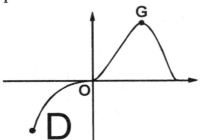

What we see is a dog where the D is its ear, and the G could be an arrow showing air flow from the dog's nose!

18. Court Out. A top is 6, B front is 5, C top is 3 and C front is 1. Since opposite faces total 7, the hidden face of C is 2. With 2 and 4 used, the blank faces of C must use 3 and one of 1 or 6 (not both since they are opposite faces). A top cannot be 5 (opposite 2) so must be one of 6 or 1 leaving B front as 5. The left of A can now be 3 or 4 but since the hidden total is even (B's hidden total is 7) then the hidden face of A is 3 and the left face is 4. For an even front total, C front is 1 or 3. If it is 3 then A top is 1 or 6 with C top as 6 or 1. However, considering the orientation of the 2, 3 and 5 on A and C, their top numbers must be impossibly identical. So C top must be 3, C front is 1 and A top is 6.

19. Having a Ball. Toddler Ball and Rollie Ball stayed dry. This is a puzzle about equal volumes and sums of cubic numbers. The volume occupied by a sphere is proportional to its diameter cubed. So if we give Baby Ball a volume of 1, then Toddler has volume 8 (2x2x2), Freda 27 (3x3x3) and so on. There are only two ways to make three cubes sum to a fourth cube using the digits 1 to 9:

$$3^3 + 4^3 + 5^3 = 6^3$$

$$1^3 + 6^3 + 8^3 = 9^3$$

The water level of the pool stays the same if the volumes occupying the pool are identical. So 3, 4 and 5 must have dived in first, were replaced by 6 (fourth cube to enter the pool) who was joined by 1 and 8, all being replaced by 9. The missing Balls were 2 and 7, that is, Toddler and Rollie.

20. Bubble Math. If straight lines are mentally added as shown we see a die (clue "Di"). One of the spots on the left face must be moved to complete the four on the top face.

21. Door to Door. The answer is 2 and 8. The 'first place' referred to is the first place of each number spelled out: **TWO**, **EIGHT, NINE** which makes **TEN**. The numbers 2 and 8 taken together get 'to equal first place'.

22. The Builder's Problem

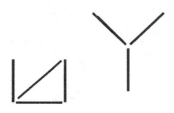

The sticks are rearranged as above and then one needs to turn one's head 45 degrees clockwise. The sand will happily sit inclined in the box.

23. Neddy's Workload. The blue cart has 16 boxes and the red has 6 boxes. Let the weights of a box of apples and oranges be a and b, respectively. Then $b = pa$ where p is a digit. Let the lengths of the blue and red carts in numbers of boxes be m and n, respectively. Then

$$m/n = (a + b)/(b - a) \qquad (1)$$

where $m > n$ and are both digits. The maximum load is expressed by the equation

$$am^2 = bn^2 + a + b \qquad (2)$$

or

$$b/a = p = (m^2 - 1)/(n^2 + 1) \qquad (3)$$

so that from (1) and (3) we have

$$(n + m)/(m - n) = (m^2 - 1)/(n^2 + 1) = p$$

The only digital solution is $m = 4$, $n = 2$, $p = 3$.

24. Moving Clocks

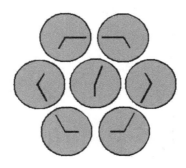

25. Armless Aliens. There were 14 Creakies and 56 Rusties. A robot with n arms requires n females and vice-versa, and the number of members is $2n$ with n^2 links. Let the Creaky tribe have x arms each and the Rusty tribe have y. Then the maximum number of links that the Rusty tribe can make after the mutilation is $y(2y-1)/2$. The number of extra arms each Creaky needs is $24y/2x$ allowing $x(x+24y/2x)$ links. Forming the links equation we have

$$2y^2 - 49y - 4x^2 = 0$$

We must have $y=4x$ so that $y-4x$ is a factor. A long division shows that the condition for zero remainder is $x=7$ so $y=28$. The numbers of members are twice these.

26. A Pressing Problem. The smallest number of button pushes is 22 for making all buttons yellow. The other minimum numbers are 23 for red, 24 for blue and 25 for green. If the special pairs did not exist the numbers would be 27 for yellow, 23 for red, 23 for blue and 23 for green. The special pairs achieve a saving of button pushes. For example, for B1 (red) controlling C3 (green), if we wanted yellow for all buttons, pressing B1 moves B1 from red to yellow and C3 from green to yellow, a saving of three pushes. Considering the savings produced by the three special pairs for each possible target colour gives the solution.

27. Amazing. The real maze is the black line. One might imagine a cut-out painted black in a white marble floor.

28. Horse Play. Clockwise from 4, the numbers are 4, 6, 2, 13, 5, 3, 10, 8, 12, 1, 9, 11 with 7 in the middle. Number 7 must be central since it is the only number appearing in six triplets totalling 21. This gives 10 opposite 4 and 2 opposite 12. For the number between 2 and 4, only (6,2,13) and (8,2,11) with 2, and (6,4,11) and (8,4,9) with 4 are possible triplets with 6,8 or 11 as candidates. If 8 is between 2 and 4, the opposite number is 6

159

which with 10 gives 5. The 8 and 2 give 11 which with 5 give 5 (invalid). If 11 is between the 2 and 4, the opposite number is 3 which with 10 gives 9, however the 11 and 2 also produce 9 (invalid). So 6 works and the solution easily follows.

29. A Raft of Changes. This puzzle originally appeared in *Puzzles for Pleasure* entitled "In the Same Boat", but the erroneous solution I presented demands an improved version of the puzzle, which I now give in full detail.

The spies are B (light) and J (heavy). The problem amounts to asking: Which two team members in the three consecutive sets of three position groups (where the initial set of three is ABCD, EFGH, IJKL) can reproduce the three corresponding raft inclinations that framed the group leader? We first need to discover the identity of the group leader, together with his weight relative to an ordinary team member (heavier or lighter), by finding the members of the three position groups after the first set of interchanges. Information is provided by an analysis of the way the two tests must cooperate in order to identify a single imposter.

Consider 12 members in three position groups of four, all having the same weight X except the imposter (group leader) who has a different weight Y. The raft has three possible inclinations: (i) forward, (ii) balance, or (iii) backward tilt. There are only three effects on the raft inclination of rearranging the team: unchanged, or a change to one of the other two chosen from (i), (ii), (iii). The group leader is necessarily determined at the end of the second test (set of interchanges) if there are no more than three suspects presented to that test so that each can be associated with a different inclination by appropriate placement.

The first test must decide between groups of suspects in order to present one suspect group to the second test. Since there are only three raft inclinations, there can be no greater than three members to a group if the second test is to be discriminating.

With the initial forward tilt of the raft, there are eight suspects for the group leader ABCDIJKL, all from the ends of the raft, which must be partitioned into groups of 2,3,3.

1. First test

The rearrangement must consist of four interchanges involving ABCEFGIL which will allow discrimination between three groups of suspects by connecting each group to a different effect.

(a) *Reversal group (2 suspects, 1 interchange)*
   If the group leader were in this group, he would reverse the forward inclination to a backward inclination when the members are moved. Here, only suspects participate in the interchange, being moved from one end of the raft to the other. This is carried out in one interchange involving two group members.

(b) *Balance group (3 suspects, 3 interchanges)*
   If the group leader were in this group, the forward inclination would become balanced when the group members were moved. Suspects are moved from the ends to the middle of the raft demanding three interchanges (leaving the reversal group with one), with three non-suspects from the middle.

(c) *Unchanged group (3 suspects, no interchanges)*
   No member of this group is moved. If the group leader were among them the raft would remain forward tilted.

These four interchanges associate each group with a different effect of rearrangement so that after noting this effect only one of these groups need be presented to the second test.

2. Second test

A rearrangement in the second test permits discrimination between either two or three suspects. The group 1(a), 1(b), or 1(c) containing the group leader has, by this time, been identified

161

after noting the effect of the first set of interchanges. We must now find out who the leader is in this group.

(a) *Two suspects*

Only the reversal group applies here. There are two ways to identify the leader.

(α) One of the two suspects is placed in the middle (Y is identified by a resulting balance) and the other at one end (Y identified by imbalance). Reference to the initial forward tilt should give the leader as heavier or lighter than the normal weight.

(β) One suspect is moved to the same end of the raft as the other. Again the leader can be identified and his relative weight obtained by noting the history of raft inclinations.

(b) *Three suspects*

These arise from the balance or unchanged groups. One suspect is placed at each end and one in the middle. Again the inclination history reveals all necessary information.

Deductions

Given information
Initial suspects: ABCDIJKL (forward incline)
Four interchanges: ABCEFGIL
Final positions: ACEI, DIJK, BFGH

Test	Back	Middle	Front	Incline
(0)	ABCD	EFGH	IJKL	forward
(1)	?	?	?	?
(2)	CFHI	DEKL	ABGJ	?

3. Unchanged group

The suspects not interchanged from (0)-(1) – see the initial suspects and the four interchanges – form the unchanged group $D_u J_u K_u$. We note that H does not move and is not a suspect.

$$(1) \qquad D_u \qquad\qquad H \qquad\qquad J_u K_u$$

Since this group is not distributed in (2) according to the test specifications 2(b) – they instead appear together in the middle – then they are not being examined in the second test and the leader is not among them.

4. Reversal groups

The possible reversal groups – see 1(a) – are those pairs that might exchange ends from (0)-(1) during the four interchanges, namely, AI, BI, CI, AL, BL, CL. By imagining their positions in (1) after juxtaposition from (0), the possibilities that might be under test in (2) are CI, AL, BL, CL – see 2(a). It is given that the leader occupies the same position in (1) and (2) which reduces these to CI (leader I), AL (leader A) or BL (leader B).

<u>Case $C_r I_r$ (leader I)</u>
The balance group deduced from the remaining suspects is $A_b B_b L_b$ which attends the middle:

$$(1) \qquad D_u I_r \qquad A_b B_b L_b H \qquad C_r J_u K_u$$

We have the given condition that the leader (which is I here) is not in a group next to either A or H so the present case is eliminated.

<u>Case $A_r L_r$ (leader A)</u>
This time the balance group is $B_b C_b I_b$:

$$(1) \qquad D_u L_r \qquad B_b C_b I_b H \qquad J_u K_u A_r$$

We eliminate again according to the leader's position.

<u>Case B$_l$ L$_r$ (leader B)</u>

This time the balance group is A$_b$C$_b$I$_b$:

(1)   D$_u$L$_r$ A$_b$C$_b$I$_b$H   J$_u$K$_u$B$_r$

We eliminate again according to the leader's position.

So no reversal group is being tested in (2) and this group does not contain the leader.

5. Balance group

The possible balance groups – see 1(b) – with the unchanged group already determined are ABI, ABL, ACI, ACL, BCI, BCL. (We cannot have ABC since one of these three is required for the reversal group.) Only the possibilities ACL and BCL could be under test in (2) – see 2(b). This gives CL in the balance group with one of A or B which must all appear in the middle in (1). So I and the remaining one of B or A occupy the reversal group and exchange ends from (0)-(1).

(1)   D$_u$I$_r$ (A/B)$_b$C$_b$L$_b$H   J$_u$K$_u$(B/A)$_r$

6. Final deductions

Let us compare (1) with (2).

(1)   D$_u$I$_r$ (A/B)$_b$C$_b$L$_b$H   J$_u$K$_u$(B/A)$_r$
(2)   CFHI   DEKL   ABGJ

If the leader sits in the balance group and occupies the same position in (1) and (2) then the leader can only be L, who must deduce that he is heavier than the normal weight (even though he isn't!) to produce the forward inclination in (0). Furthermore, A cannot be at the front in (1) otherwise he would be in the next group to the leader L (invalid) and so A must be in the middle with B at the front.

(1)   D$_u$I$_r$ A$_b$C$_b$HL$_b$ BJ$_u$K$_u$
(2)   CFH<u>I</u>   DEK<u>L</u>   A<u>B</u>G<u>J</u>

Note that from (1)-(2), BIJL do not move (underlined). From (1), one interchange must be A in the middle with K at the front. Both C and H in the middle in (1) must go to the back in (2) by interchanging with D and E both of which must be at the back in (1). This exhausts the 3 interchanges and so F must be at the back and G at the front in both (1) and (2). So we get:

(1) $D_u EFI_r$ $A_b C_b HL_b$ $B_r GJ_u K_u$

(2)   CFH$\underline{I}$ DEK$\underline{L}$ A$\underline{B}$G$\underline{J}$

We know that for the two enemy spies (one lighter by the same amount as the other one is heavier than the chosen weight) to produce balance in (1) and (2) they must stay together. This allows CH, FI, DE, BG, BJ, GJ for the possible pairs of suspects and they must produce the raft inclination of two men. One must be at the back (lighter) and the other at the front (heavier) since it is given that the inclination in (0) was "that which might be expected from two men having a different weight from the rest". This is only possible for B (lighter) and J (heavier).

30. Right Angle. A quarter turn of one's head anticlockwise reveals the name MEXICO when reading down the new letters.

31. Square Feet with Corn. Rastus received 9 parts and the triplets each got 13 parts. Let the work rates of the triplets combined, Rastus and Willy be x, y, z, respectively. Then

$$5(x + y) = 6(x + z)$$

and we have $y = (x + 6z)/5$. Since z has 1 part we can let $x = nz$ where n is a whole number. We then have $y = (6 + n)z/5$. The ratio of the work rates of x, y, z is now $n : (6 + n)/5 : 1$. Let the total number of parts be m, a two-digit square number. We then have

$$n + (6 + n)/5 + 1 = m$$

so that $n = (5m - 11)/6$. The only solutions are $n = 19$, $m = 25$ and $n = 39$, $m = 49$. Since n is the total number of parts the triplets receive, only the latter value of n is suitable (divisible by three).

32. Nothing for It. Place the N to the left and the R to the right. The N can be rotated a quarter turn to make a Z. We can then have ZERO = ZERO.

33. The Tin Door

The reason that the door is made of tin is so that the reflection of the message on the wall can be seen in it. The message is "tin mine inside it".

34. Play on Words. Sally's chosen strategy is a losing one. The only restrictions on the game are that 'afraid' is not chosen first and that Sally intends to choose it on her second turn as the third selected envelope should it be available. In that case, 'fear', 'alarm' or 'frightened' could follow it as the fourth and last envelope with equal probability 1/3. This gives her opponent one chance to take 'frightened' which is four letters more (+4) than 'afraid', one chance to get 'fear' (−2) and one chance to obtain 'alarm' (−1). So on the fourth envelope, the expected gain for Sid (or loss for Sally) is 4/3–2/3–1/3 (that is, the sum of the products of gain and probability) which is 1/3. We multiply this by the number of times the game is run for Sally's total loss of letters due to her strategy. It would be a better strategy for Sally if she only took 'afraid' on her second turn if she had taken 'frightened' on her first one.

35. Doubtful Date. The girl is 18 years old. Write out the 17 and partition the letters as follows:

WEDS/EVEN/TEEN

So the only age that is at least 18, an even number, and a teen age is 18.

36. Shilly Chalet. The rooming arrangements were as follows. Neil and Ian were in chalet 1, Martin and Colin in chalet 2, and Keith and Gordon in chalet 3.

37. Pet Theory

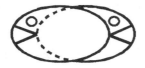

Add the lines as shown to produce two fish, one hiding behind the other. Note that the tail (on the right of the picture) of the left-facing fish exactly covers the open mouth of the right-facing one behind it, while the open mouth of the left-facing fish (on the left of the picture) is exactly the same size as the tail of the right-facing fish.

38. The Broken Ruler. There are 72 boxes in the smallest possible packing case volume of 1728 cubic inches. If we assume that the rule shows a dimension of one inch then this implies that the three chocolate box dimensions are 1×2×3. Let the number of chocolate boxes along the packing case edges be x, y, and z. Then the volume of the packing case is the product $6xyz$ cubic inches. If this volume is to be a cube number then we must have $xyz = 2^{3m-1}3^{3n-1}$ where $m, n = 1, 2, 3, \ldots$ The smallest non-square number for xyz occurs for $m = 2$, $n = 1$ giving $xyz = 288$ with a packing case volume of 1728 cubic inches. However, xyz violates the condition that it is to have two digits. There is another way to interpret the picture of the broken rule in that it appears inverted with a missing 7 and that the digits shown are really 6 and 8. This gives one chocolate box dimension as 2 inches allowing the three dimensions to be 2×3×4. The packing case then has volume $24xyz$ and $xyz = 2^{3m}3^{3n-1}$. The smallest non-square xyz occurs for $m = 1$, $n = 1$ so that $xyz = 72$. This is now two digits and again gives the packing case volume as 1728 cubic inches. What we notice about this solution is that

167

it satisfies another interpretation of the given question in that the number of chocolate boxes 72 "can be found" in the number of cubic inches that constitutes the packing case volume 1728.

39. The Four Dice. The missing number is 2. Each die is actually a digit: 6, 5, 9, 3, which denotes the sum of the two visible faces on that die.

40. Back to Class. Droopy had 8, Dimwit had 4 and Dibdib had 9. Let the first three digits be A, B, and C. The third statement gives

$$10A + B - (10B + A) = 3(A + B)$$

so that A=2B for the first two digits. Combined with the first statement, which deals with the last two digits, we can then have 216, 425 or 849. The second statement then allows only the last alternative.

41. Fare Enough. Only Swigger and the driver made a profit of £2.50 and £2, respectively. The cost to Hughie's house was 8×£1.80=£14.40. Divided by three, each person pays £4.80. The cost of the next mile, to Swigger's house, is shared by Swigger and Ralph, that is, £0.90 each. The cost of the final mile is met by Ralph who pays £1.80. So Hughie should pay £4.80, Swigger £5.70 and Ralph £7.50.

Hughie gave up £5 (to Swigger), Swigger gave up £10 (£20 to Ralph minus £5 from each of Ralph and Hughie), and Ralph gave up £5 (to Swigger). The driver received £20 (£18 fare plus £2 tip).

So Hughie lost £0.20, Swigger lost £4.30, Ralph gained £2.50 and the driver gained £2.

42. No Escape. Moving the door knob as shown turns the right-hand door frame into an open door with the edge of it facing us.

The original position of the door knob causes us to expect it to swing open to the right. The letterbox becomes a brick revealing that when the door is opened there is a brick wall behind it.

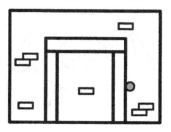

43. The Baffled Brewer. We have 3 from C to A, 2 from B to A and 1 from C to B with the definitions of A, B and C as below. The total beer is 33 gallons so 11 litres begin in each barrel. Let the barrel that is to contain 16 litres be A, 10 litres be B and 7 litres be C. So A must gain 5 litres, B loses 1 litre and C loses 4 litres.

Since each jug is used once only and each barrel must be involved in at least two transfer operations, then we must have exactly two operations for each barrel. This means that A must have +3 and +2. There are two ways that C can lose 4: −2, −2 or −3, −1. For he first case, the 2 litres jug is used twice which is invalid. For the second case, B has −2,+1. This leads to the solution.

44. Sound Arithmetic. He used nine 3 honk, two 4 honk and four 7 honk coins. Let the numbers be x, y, z, respectively. Then

$$3x + 4y + 7z = 63$$

$$x + y + z = 15$$

Multiplying the second equation by three and subtracting the result from the first gives

$$y + 4z = 18$$

with the solutions $y=2$, $z=4$ or $y=6$, $z=3$. The latter gives $x=6$ (invalid since $x=y$) so the former holds with $x=9$.

45. The Pig and the Bird

The bird is facing us with head bowed, the pig's ears become its wings, the neck forms its legs, the nostrils become its eyes, the snout its head, and the crosses show that the bird is bruised. The added triangle forms the beak.

46. Out of this World. Rotate the picture 90 degrees anticlockwise and add ET to spell ROCKET.

47. Doing a Turn. The answer is 2. The question is asking about the identity of Q! Rotate the diagram a quarter turn anticlockwise to find three mathematical statements about Q. On the right of each statement is a rotated number. It is given that the solution is a "whole number". The first statement is that it is less than 4, the second that it is greater than 1, and the third that it is not equal to 3. This only leaves 2.

48. The Concealed Car. Fully open door D onto door C.

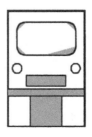

49. The Dead Dog. The front leg must be moved to the top pointing up. We then see a dog lying on its back with its legs in the air, the back leg has become a horizontal tail, the tail has become the back leg, the mouth becomes the eye, and the eye becomes the mouth.

50. Water Puzzle. If the stick is placed horizontally at the top of one of the glass stems the letters H-TWO-O are visible, the W being the two curves of the wine glasses, and the H arising from a rotation of the constructed letter. H_2O is the chemical formula for water. Roger Procter from Tanzania has suggested an inversion of the picture to show two taps, with a vertically placed stick being running water from one of them.

51. Arithmystic. Spell out the numbers: ONE + NINE-TYEIGHT − TEN − EIGHTYNINE = O, where the last character is the letter O not a zero 0. However, the following sets of letters are not equivalent: ONE + NINETYEIGHT and TEN + EIGHTYNINE.

52. The Slug and the Snail. The snail reached the finish in 5 seconds. Since they reach the same position D at the same time t we have

$$2V(t-T) = X + Vt = D$$

Rearrangement gives $V(t-2T) = X$. Any solution where $t = 2T + 1$ cannot work since then $V = X$ (must be different numbers). For the set of solutions where $T=1$; $t=4, 5, 6$; and $D=6V, 8V, 10V$, respectively, only $V=1$ allows $D < 10$ but then $V=T=1$ (invalid). For $T=2$, $t=6,7$ and $D=8V, 10V$, respectively. Again $V=1$ which gives $X=2$ for $t=6$ ($T=X$ invalid) and $X=3$ for $t=7$ with $D=10$, a valid solution. For $T=3$, $t=8$ and $D=10V$. Here $V=1$ gives $X=2$ with $D=10$, another valid solution. So there are two valid solutions, both with $V=1$ and $D=10$, with $T=2$, $t=7$ and $T=3$, $t=8$. In both cases $t - T = 5$.

53. Fish Feast, The route was XECDBAX as shown resulting in the fish at X becoming enlarged after eating the others (with the letter O as its eye)! The other possible route XBCDEAX, showing the fish swimming to the right is invalid since the direction is not *with* the shoal.

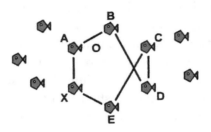

54. Secret City. Remove the right-hand leg of the letter M in MY to give NY CITY, that is, "New York City".

55. No Earthly Connection. Turn the W upside down to make a light bulb. This is connected to the live (L) and the neutral (N) but not to the earth (E) which has broken connections.

56. Inspector Lewis. The table and mug are constructed from the letters A,L,I,C,E. Note the clues: Lewis, Karel (Carroll), looking, and glass.

57. Inklined. Looking at the bottle very closely, it is constructed from the digits 2 and 5 which represent the filling time 25 seconds. If the bottle is now tipped upside down, thereby allowing the bottle to empty, we see that these digits look like 5 and 2 and so the bottle empties in 52 seconds.

58. Miserable Marriage. The letter B is constructed from two letter D, one on top of the other. Remove the top D to leave DAD, where the middle letter A can be interpreted as steps, so step Dad!

59. The Backward Robber. The storekeeper got 3375 coins. If the robber hands over one half, can count out one third of the remainder (a half) and hand over one half of the remainder then the original amount is a square whole number divisible by 2 × 2 × 3=12. So the three-digit candidates are 2 × 2 × 2 × 2 × 3 × 3 = 144, 2 × 2 × 3 × 3 × 3 × 3 = 324, 2 × 2 × 2 × 2 × 2 × 2 × 3 × 3 = 576 and 2 × 2 × 3 × 3 × 5 × 5 = 900. Since the robber actually handed over three quarters of this number five times to give a cubic number, the only possibility is 900. The storekeeper received five lots of three quarters of 900.

60. Dig It. Pick up (remove) the stone and draw a horizontal line on the spade blade. Viewed upside down, the weight of the stone on the spade handle has driven it deeper into the ground.

Finger Multiplication. Here is an explanation for why it works. If we take our earlier calculation for 9×8 with the eight times table we can rewrite it as follows:

$$9×8 = 9×(10- 2) = 90- 18 =$$
$$90- 20 + 20- 18 = (9- 2)×10 + 2 = 72$$

Notice that the jumps of two appear with (10- 2) and our nine jumps of two (18) is counted on the fingers. When this is subtracted from the next highest ten (20) we get the units digit (2) and when the number of the pass is subtracted from the position (9) in the multiplication table (9- 2) we get the tens digit (7).

I have observed that a bright eleven year old can master this method in half an hour.

About the Author

International puzzle master Barry R. Clarke writes math and logic puzzles for *The Daily Telegraph* (UK) and has contributed enigmas to many other publications including *New Scientist*, *The Sunday Times* (UK), and *Brain Games* (USA). He has acted as consultant for BBC TV's *Mind Games* and his books include the best-selling *Brain Busters* (Dover: 2003) and Mensa's *Challenging Logic Puzzles* (Sterling: 2003). He has also published academic papers in quantum mechanics and Shakespeare studies.

· 13 · 12 · 12 · DU · Γ3 · π : ·

Produced by Quarto Anti-rota-type